Wolfgang Steinicke

Nebulae Star Clusters Galaxies

History – Astrophysics – Observation

Front cover: The planetary nebula M 97 (below left) and the galaxy M 108 in Ursa Major
Back cover: The bright galaxy NGC 7331 and the galaxy group Stephan's Quintet (below right) in Pegasus

© 2019 Wolfgang Steinicke

This publication is in copyright. Subject to statutory exception and to the provisions of relevant collective licensing agreements, no reproduction of any part may take place without the written permission of the author.

ISBN 9783749451784

Manufacturing and publishing: BoD- Books on Demand, Norderstedt, Germany

To my wife Gisela

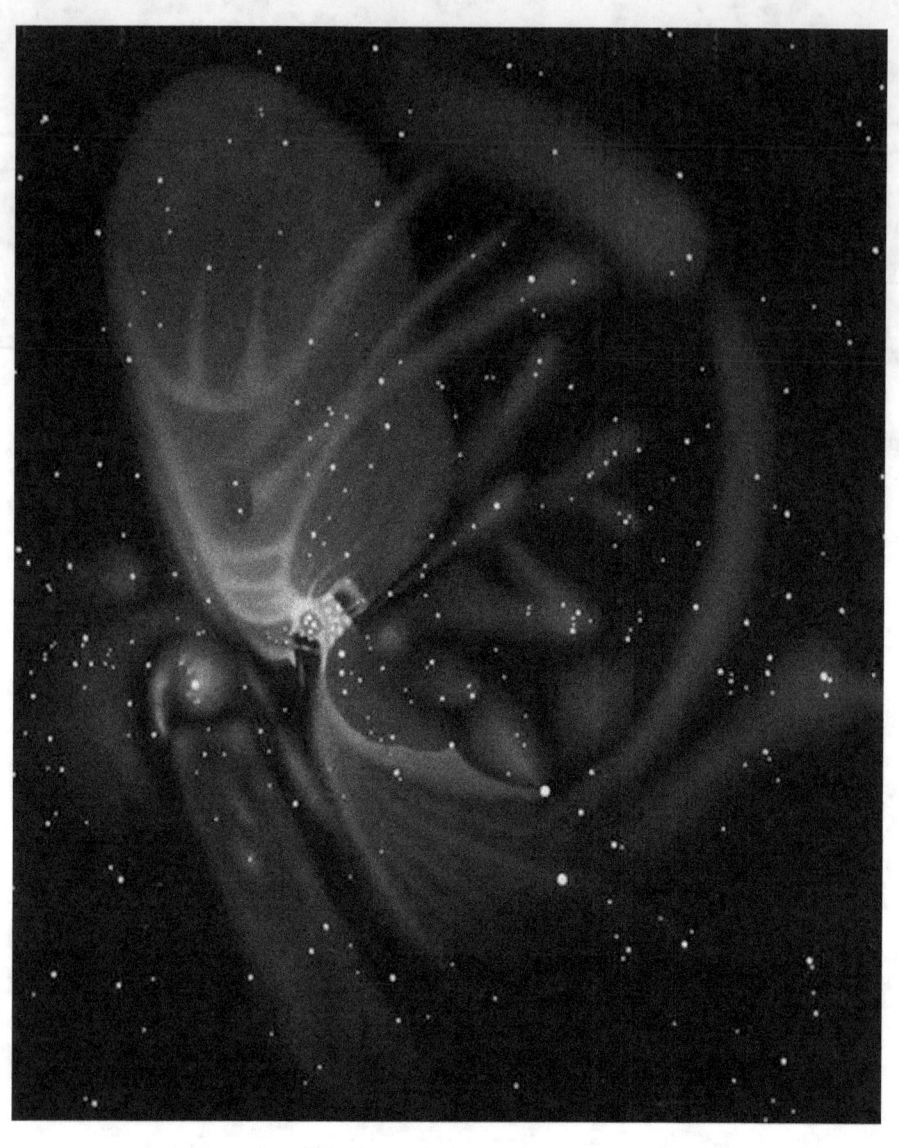

Orion Nebula (drawing by Wilhelm Tempel, 1877)

Table of Contents

INTRODUCTION   1

HISTORY OF DEEP-SKY OBSERVATION   3

Early Discoveries and the Messier Catalogue   4

The Epochal Work of William, Caroline, and John Herschel   9

First Ideas about Star Clusters and Nebulae   17

Lord Rosse's Spiral Nebulae   19

Spectroscopy, Photography and the True Nature of Galaxies   24

OBJECT TYPES AND THEIR ASTROPHYSICAL NATURE   29

General Data of Deep-sky Objects   30
    Celestial Position and Orientation   30
    Brightness, Magnitude, and Size   32
    Data Sources and Catalogues   33

Open Clusters   39
    Physical Nature   39
    Classification   43
    Associations, Moving Groups, and Asterisms   44

Globular Clusters   45
    Structure and Age   46
    Distribution and Classification   47
    Extragalactic Globular Clusters   48

Diffuse Nebulae   48
    Emission Nebulae, Supernova Remnants   49
    Reflection and Dark Nebulae   50
    Bipolar and Cometary Nebulae   52

Planetary Nebulae   53
    Distribution and Physical Nature   53
    Spectrum and Classification   54

The Milky Way   55

Galaxies	**60**
Brightness, Size, and Orientation	60
Classification	62
Distance	69
Diameter, Luminosity, Mass, and Rotation	70
Active Galactic Nuclei and Quasars	72
Pairs and Groups of Galaxies	74
Clusters of Galaxies	76
The Evolution of Galaxies and Clusters of Galaxies	76

PRACTICE OF OBSERVATION 83

Object Selection 85

Telescope and Equipment 86

Observation Site and Atmospheric Conditions 88

Object finding 91

The Technique of Visual Observing 92

Subjectivity, Description, and Drawing 94

Astrophotography	**95**
Imaging vs. Visual Observing	96
Camera Types	97

COLLECTION OF INTERESTING DEEP-SKY OBJECTS 99

Open and Globular Clusters 100

Diffuse Nebulae 109

Planetary Nebulae 116

Galaxies, Quasars, and Galaxy Clusters 120

APPENDIX 137

Bibliography, General Index, Object Index, Source of Figures

Introduction

People, when looking at the night sky, are fascinated by the Moon, planets and stars. The latter were historically arranged in constellations, like Orion or Cygnus. Perhaps one stumbles over conspicuous ensembles of stars, like the Hyades or Pleiades, located in Taurus. But the naked eye can discover more than that. There are the nebulous spots, located in Cancer, Perseus and Andromeda. Binoculars or a small telescope will easily show that the first two, known as Praesepe and χ Persei, consist of faint stars. However, the third 'nebula' stabbornly resists visual resolution – in any amateur telescope. Your view has fallen on a galaxy. These fascinating experiences may trigger a career as a deep-sky observer.

The term 'deep sky' refers to objects beyond our solar system. These objects include stars, star clusters, diffuse nebulae and galaxies. Deep-sky objects are in many ways' attractive targets – for professional and amateur astronomers. They lead the observer to great distances, and at the same time the view goes far into the past. The light of the most distant objects (quasars) has taken billions of years to reach us. But also, nearby ones, located in the Milky Way, are fascinating. We speak of open clusters and galactic nebulae.

However, one class of deep-sky objects is ignored in this book: stars. Instead, the focus is on sources of light, showing a more or less extended structure. Consequently, star clusters, galactic nebulae and galaxies are called 'nonstellar' objects. They are interesting targets for both visual observing and photography.

The book covers three major subjects: history, astrophysics and observation. The first chapter treats the important role, clusters and nebulae played in the history of astronomy. This ranges from the early times of pure visual observing, over speculations about their nature to modern astrophysics. Only in the early twentieth century with the adavancment of observational techniques such as photography and the understanding of physical processes, could reveal the true nature of deep-sky objects. It has been a long road from the Greek astronomers like Hipparchus, via great observers like William Herschel or Lord Rosse, to eminent astrophysicists like Edwin Hubble. Thanks to such great scientists, we now know a lot about the creation and evolution of clusters, nebulae and galaxies, and their place in the hierarchical structure of the cosmos – solely by observing from Earth. It is a remarkable fact that the entire information exclusively comes from radiation of various kinds, emitted by remote, unreachable sources.

The second chapter presents the various types and subtypes of deep-sky objects and discusses their astrophysical nature, based on essential quantities like distance, brightness, or size. The relevant classification schemes, data sources, and catalouges are mentioned too. A special focus is on galaxies, the building blocks of our universe. They consist of stars and interstellar matter – and a supermassive black hole in the centre. Galaxies come in various forms and tend to build pairs, groups and clusters. The largest aggregates are superclusters, marking the endpoint of cosmic hierarchy.

The third chapter is dedicated to the practice of observation. This covers instrumental factors (e.g. telescope, eyepieces, filters) and important quantities (e.g. contrast, field of view, magnification). Moreover, the methods and conditions for successful observing are discussed, like viewing techniques, based on the functions of the eye, or atmospheric conditions at the observing site. Also important are object selection and finding methods. Not only visual observing and drawing is treated, but also the important field of astrophotography.

The final chapter presents a selection of interesting deep-sky objects (along with their data), covering the relevant types. Not only easy targets for smaller telescopes are chosen, but also tough ones for large apertures or digital cameras. Following these lines, one gets a deep impression of cosmic hierarchy.

Regarding the writing of this book, I would like to thank Peter Morris (formerly at the Science Museum, London) for interesting discussions. Helpful support came from two colleagues at the Webb Deep-Sky Society, Owen Brazell and Stewart Moore. They have carefully checked my English text. Finally, I have to thank my German friend Stefan Binnewies for using some of his professional astrophotos. Though presented in black and white, they have lost none of their extraordinary beauty.

Wolfgang Steinicke, April 2019

History of Deep-Sky Observation

Early Discoveries and the Messier Catalogue

For millennia, humans must have looked up at the sky. In the pristine darkness of prehistoric times, the Milky Way with its striking mix of bright and dark clouds would have left a profound impression on those early observers. They probably resorted to mythical explanations such as the Ancient Greek story about the origin of the shimmering band. It is based on the legend that Zeus had his son Heracles, whom the mortal Alcmene had given him, to drink at the breast of his sleeping wife Hera. This was to equip Heracles with divine powers. The young hero sucked so strongly that Hera woke up. Abruptly, she pushed away the strange baby and a ray of her milk spilled over the sky. The Greek coined the name γαλαξίας (galaxías), which means 'milky nebula of stars'.

Beside the luminous Milky Way, other 'milky' spots or clouds are visible with the naked eye. There is no doubt that some of them – now identified as nebulae or star clusters – were already known in prehistoric times, although there is no record. The two Magellanic Clouds are particularly prominent in the southern sky. Alas, there are no historical representations, like rock drawings by native Australians.

The first records about extended (non-stellar) objects are due to the Greek natural philosophers. The cluster below Sirius, known as M 41, was seen by Aristotle about 325 BC, Preasepe (M 44) in Cancer by Aratus about 260 BC, and the Double Cluster in Perseus (χ Persei) by Hipparchus about 130 BC. Later, in about 130 AD, Ptolemy added M 7 in lower Scorpius. He catalogued such objects in his famous book, the *Almagest*.

Ptolemy's Cluster M 7 in Scorpius is the most southern Messier object.

The next deep-sky object recorded is due to the Persian astronomer Abd ar-Rahman Al-Sufi: of the Andromeda Nebula (M 31), seen before 964 from Isfahan and plotted as 'little cloud' in his famous *Book of the Fixed Stars*. Al-Sufi also described the o Velorum Cluster and the Large Magellanic Cloud (LMC), called 'white ox' by him. However, the Small Magellanic Cloud (SMC) was too far south for Al-Sufi to see. This conspicuous object was first mentioned by Amerigo Vespucci (together with the large cloud). The Florentine explorer saw it on his cruise in 1501, 20 years earlier than the Portugese Fernão de Magalhães.

The o Velorum Cluster (IC 2391), visible to the naked eye, was first mentioned by the Persian astronomer Al-Sufi.

No other deep-sky object was recorded until the invention of the telescope in 1608. On 24 November 1610, Nicolas Claude Fabri de Peiresc discovered the Orion Nebula (M 42) with a small Galilean telescope. Independently, Johann Baptist Cysat saw the object a year later. The Andromeda Nebula was rediscovered on 15 December 1612 by the German astronomer Simon Marius, not knowing of Al-Sufi's work. Whilst, in later years, some nebulous objects appeared in the telescope as star clusters, both the Andromeda Nebula and Orion Nebula could not be resolved.

Real progress was made by Giovanni Battista Hodierna's observations, made in about 1654 with a small refractor. He discovered 12 deep-sky objects, among them the Lagoon Nebula (M 8) in Sagittarius and the Triangulum Nebula (M 33); the first galaxy, exclusively found with the aid of the telescope. Larger telescopes were built in the seventeenth century and the list of objects grew. In 1665, the German astronomer Abraham Ihle discovered the first globular cluster, M 22 in Sagittarius. Then the famous English astronomer Edmond Halley entered the scene. His target was the southern sky, surveyed from the island of St Helena. There he discovered the brightest globular cluster, ω Centauri, in 1677. While measuring star positions at Greenwich in 1690, another Englishman, the first Astronomer Royal, John Flamsteed, found the open cluster NGC 2244 around the star 12 Monocerotis.

The first half of the eighteenth century brought about a slow increase in the number of objects found, but in the second half the rate of discovery became inflationary. 42 nebulae and clusters were discovered until Messier entered the stage in 1758. In 1714, Halley saw the bright globular cluster M 13 in Hercules. Another spectacular object was found by the Englisch amateur John Bevis in 1731: M 1, the Crab Nebula in Taurus. And in 1749, the Andromeda Nebula got a companion: M 32, found by Guillaume Legentil. The main discoverers of that period where Nicolas-Louis de Lacaille (27 objects) and Jean-Philippe Loys de Chéseaux (8 objects). The latter observed from Paris, finding the Omega Nebula (M 17) in Sagittarius (1745). Following in Halley's footsteps, Lacaille surveyed the southern sky from Cape Town in 1751, using a refractor of only 12 mm aperture. His discovery of the bright galaxy M 83 in Hydra with such a tiny instrument was exceptional.

The 'grand design' spiral galaxy M 83 in Hydra.

When the French astronomer Charles Messier, at the end of August 1758, was searching for the reappearing comet Halley with a 3.5-inch refractor, he noticed a nebulous spot 1.2° northwest of the star ζ Tauri. He first regarded it as the object sought but noted: 'Nebula without stars above the southern horn of the bull. It is of pale white light and has an oval shape, like the flame of a candle.' The true comet, however, was about 10° southeast. Later observations revealed no motion. Thus, the obscure object was shown to be a nebula, like that in Orion or Andromeda. To avoid future confusion, the famous comet hunter documented similar cases. A first list of 1771 contained 45 nebulae and star clusters, sorted by discovery date. Messier soon realized that his alleged 'comet Halley' was already discovered by Bevis in 1731. Thus, the 8.4 mag bright and 4' large object became the first entry of the list: M 1. The last one is M 45, the Pleiades in Taurus. Though this open cluster, and also Praesepe (M 44), can never be confused with a 'comet', Messier included them for the sake of completeness. 18 of the 45 objects were discovered by him, among them the Trifid Nebula (M 20) in Sagittarius and the Dumbbell Nebula (M 27) in Vulpecula.

The Crab Nebula M 1 in Taurus was visited by Saturn in the year 2003.

In 1780, Messier published an update of his catalogue, now containing 70 entries (M 70 is a globular cluster in Sagittarius). Among the new discoveries is the famous Ring Nebula in Lyra (M 57). Messier saw the bright planetary on 31 January 1779. On 13 October 1773, he found the Whirlpool Nebula (M 51) in Canes Venatici, which got its name later after the detection of spiral structure by Lord Rosse.

Messier's final catalogue, listing 103 objects, eventually appeared in 1781. It contains discoveries made until April that year. 18 were contributed by a new observer, Pierre Méchain, a close friend. Among them was M 97, the Owl Nebula in Ursa Major, seen in February 1781. Messier himself added seven discoveries. Among the new objects were the galaxies M 81 and M 82 in Ursa Major. The popular pair was found by the Berlin astronomer Johann Elert Bode in 1774, thus the name Bode's Nebulae.

When Messier – always hunting for comets – inspected the region of Virgo and Coma Berenices in 1781, he noticed a remarkable accumulation of nebulae; a few of which had already been catalogued by him. He wrote: *'The constellation Virgo and especially the northern wing is one of the constellations which enclose the most nebulae. The catalogue contains 13 which have been determined. All these nebulae appear to be without stars and can be seen only in a good sky and near meridian passage. Most of these nebulae have been pointed out to me by M. Méchain.'* Messier had found the Virgo Cluster, the nearest cluster of galaxies.

In the final catalogue, 41 of the 103 objects must be credited to Messier and 19 to Méchain, followed by Hodierna (8), Koehler and de Chéseaux (6 each). The rest is due to 18 different observers. In the twentieth century, the original Messier catalogue was enhanced to 110 entries. The additional objects are based on unpublished observations of Méchain (M 104–109) and Messier (M 110); the latter is the second companion galaxy (NGC 205) of the Andromeda Nebula.

There is no doubt, the Messier catalogue collects many of the finest deep-sky objects of the northern sky. Looking at the modern version with 110 entries, we have 40 galaxies, 29 globular clusters, 27 open clusters, 7 diffuse nebulae (including the supernova remnant M 1) and 4 planetaries. What about the remaining seven? Three objects are special (all found by Messier): M 24 is a large star cloud in Sagittarius, M 40 in Ursa Major is merely an optical pair of stars and M 73 in Aquarius is a small random ensemble of four stars. Moreover, there are four 'lost objects': M 47, M 48, M 91 and M 102. Due to poor information in the orginal catalogue, they could not be identified with existing objects. However, recent investigations brought the answer: M 47 and M 48 are star clusters (Puppis, Hydra), M 91 and M 102 are galaxies (Coma Berenices, Draco).

Messier's object M 73 in Aquarius (found in 1780) is only a random ensemble of four stars (10 –11 mag, diameter 1.5').

The Epochal Work of William, Caroline, and John Herschel

Messier's work was continued by William Herschel – no doubt, the greatest visual observer of all time. His surveys of the northern sky, made from small English villages near Windsor Castle and lasting over 30 years, rocketed the number of known deep-sky objects from 110 to about 2,500!

Astronomical giants: William Herschel (1738–1822), his sister Caroline Herschel (1750–1848), and his son John Herschel (1792–1871).

The German born astronomer started his observational career about 1774 and his first target was double stars. Detecting unexpected cases like Polaris and Castor, his ambition for the deep-sky was awakened. However, he never neglected to observe solar system objects. Herschel performed so called 'star reviews', mainly a search for double stars, using a fine 6.2-inch Newtonian reflector and the best available data, in form of Flamsteed's *British Catalogue* of about 3,000 stars and the corresponding *Atlas Coelestis*. Equipped with these tools, Herschel systematically inspected bright stars and their vicinity. Until the end of September 1783, when the third review was finished, he had discovered about 700 double or multiple stars.

An epochal by-product of this campaign was the discovery of Uranus on 13 March 1781, then in Taurus. This brought him great recognition by his colleagues – and, of course, King George III, an admirer of astronomy. As a consequence, Herschel moved from Bath to Datchet in the Windsor area in 1782 to become the King's personal astromomer. In the following years, He often was visited by members of the royal court, eager to look through his unrivalled telescopes – stealing from the master precious observing time.

Occasionally, Herschel encountered nebulae and star clusters in that period. His first find was the open cluster NGC 2232 in Monocerotis, seen on 5 December 1779; 10 more non-stellar objects should follow until September 1783. The most spectacular was the planetary nebula NGC 7009 in Aquarius, discovered on 7 September 1782 and later called Saturn Nebula by Lord Rosse due to its peculiar appearance.

Willam Herschel's first deep-sky discovery was made with a 6.2-inch reflector in 1779: the open cluster NGC 2232 in Monoceros.

During the star reviews, Herschel also found time to look-up deep-sky objects, listed in his copy of Messier's catalogue. He observed 68 of the 103 M-numbers, using selfmade metal-mirror (speculum) reflectors of up to 12 inches aperture. Herschel was proud that his instruments were superior to that of Messier or any other astronomer of the time. The Frenchman had described many objects as 'nebula without stars', i.e. he was unable to resolve them in his 3.5-inch refractor. As the leading maker of large reflectors, Herschel wanted a competition – and Messier's objects were the ideal targets. Concerning resolution, he was successful in many cases where Messier had failed. Herschel also discovered 'garnet stars', showing a deep red colour. The most famous is µ Cephei, first seen on 27 September 1782 and known as Herschel's Garnet Star. Another field of research was the monitoring of variable stars.

Some small, round nebulae turned out to be spherical aggregations of large numbers of stars, called 'globular clusters' by him. They offer a certain range of concentration. In compact cases, like M 80 in Scorpius, he could only perceive faint stars in the outer region. But there were also loose exemplars, like M 55 in Sagittarius, where even the core appeared resolved. Other nebulae, like M 31 in Andromeda, showed no stars even with the highest magnification. It should be noted that Herschel used powers up to incredible values of about 5000, which caused incredulous amazement among contemporary astronomers. During his Messier survey, Herschel encountered also mixed cases, where stars are embedded in nebulous matter. For instance, the Orion Nebula has a close quartet of stars in its centre (θ Orionis), called 'Trapezium' by him.

The successful observations of Messier objects and the unexpected discovery of some new ones inspired Herschel in 1783 to change the observational field from chasing after double stars to the search for nebulae and star clusters.

However, there was a second stimulus, coming from his able sister Caroline. In late August 1782, she started own observations with a small refractor, given to her by William. However, this was not voluntary: 'I found to be trained for an assistant astronomer.' She should scan the sky for all kinds of objects (expecting known stars, clusters and nebulae from the Messier catalogue) – or perhaps comets. However, her brother was always near to assist or check promising finds. Until the end of October 1783, Caroline had used 65 nights. Mostly, William was observing too, working for the star review, inspecting Messier objects, visiting Uranus or other planets. Caroline's refractor was replaced in May 1783 by the 'small sweeper', a fine 4.2-inch Newtonian reflector with 2 feet focal length and 2.2° field of view. In 1793, she could even upgrade to a 9.6-inch ('large sweeper').

Finally, Caroline has viewed 52 Messier objects. But there was still greater yield: 9 new open clusters, a galaxy and – what brought the greatest honor – 8 comets! Her first deep-sky discovery was the open cluster NGC 2360 in Canis Major, proudly noting 'Messier has it not'. On 23 September 1783, the galaxy NGC 253 in Sculptor was found. Watching Caroline's success, William was impressed – and a bit jealous. It convinced him that there still was much to discover in the virgin field of nebulae and clusters. Obviously, the former observations by astronomers, leading to the Messier catalogue, had missed many objects. Thus, he summarized the plan for a comprehensive sky survey with the aid of a superior telescope.

Sketches of the nebula NGC 253 in Sculptor made in 1783 by Caroline (left) and William Herschel.

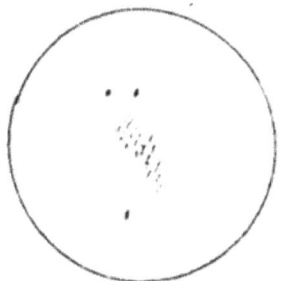 

On 23 October 1783, Herschel started his great campaign to sweep the entire sky, visible from southern England, for new nebulae and star clusters. At the pretty dark site (Datchet), the declination limit was about –30°. The instrument, constructed for the ambitious task, was a Newtonian reflector of 18.7 inches aperture and 20 feet focal length, held by a wooden altazimuthal mounting. It became operational in September 1783 and was large enough to see even faint

nebulae, but not too cumbersome to allow effective visual observing. The standard eyepiece had a magnification of 157, offering 15' field of view.

William Herschel's standard telescope for sweeping: the 18.7-inch reflector.

Herschel developed a sophisticated method called 'sweeping'. The reflector was fixed in azimuth to the south meridian. At the chosen elevation, the sky passed through the field of view from east to west due to the Earth's rotation. By slowly swinging the tube up and down by about 2° over many hours, a large rectangular area of the sky was covered. In October 1786, the Newtonian design was given up by removing the secondary mirror. By slightly tilting the main mirror, the focus went to the edge of the tube opening, where the eyepiece was now placed. The new design, called 'front view', meant he could see objects about half a magnitude fainter.

Herschel perfomed no less than 1112 sweeps in 616 nights. No minute was left out when there was a chance for clear sky. During the day, he was mainly employed with telescope making. Obviously, this great man of exemplary persistence and power did not need much sleep. The result was overwhelming: from 1783 to 1802, Herschel discovered about 2,500 nebulae and star clusters. Moreover, in 1787, he found two moons of Uranus with the 18.7-inch: Titania and Oberon.

Of course, this enormous work was impossible without careful help – and the right person was nearby: William's talented sister Caroline, perhaps one of the most undervalued persons in the history of astronomy. She was responsible for important tasks on the sweep campaign (with the exception of the observing itself, along with certain technical matters). Caroline planned the observations (selecting sky area and possible targets), documented them in real time (sitting in a room near to the telescope) and reduced the data on the next day. By recording time and elevation for each object crossing the meridian, she calculated relative positions to nearby Flamsteed stars. For this, she was well trained in mathematics by her brother. The new objects were numbered and later classified. Caroline also calculated coordinates: right ascension and 'polar distance' (= 90° − declination). For instance, a star at the local zenith (51° declination) has 39° polar distance.

Caroline finally produced the three Herschel catalogues of nebulae and star clusters, published 1786, 1789 and 1802 in the *Philosophical Transactions of the Royal Society*. They contain 1000, 1000 and 500 entries, respectively. In addition, based on William's observations, she revised Flamsteed's *British Catalogue*. The result was published in 1789 by the Royal Society under her name. Finally, she compiled the (unpublished) *Zone Catalogue*, arrangeing all 2,500 Herschel objects in zones of polar distance. This work was done for her nephew John. John Herschel wanted it for his ambituous task to confirm and enhance his father's data by a new observing campaign. In 1828 Caroline's work was honoured with the Gold Medal of the Royal Astronomical Society. This institution was founded in 1820 in London and William Herschel became its first President (two years before his death).

The many documents and records, written by William and Caroline Herschel, are archived at the Royal Society and the Royal Astronomical Society. From them the sweep campaign was reconstructed by the author. This includes the observing program, telescope configuration, sweeping method, data recording, reduction and compilation up to the printed catalogues and later publications. One major result is Herschel's sky coverage. Due to the overlapping sweep areas, 90% of the sky, visible from the Windsor area, was theoretically observed. However, the sweep pattern (by the two perpendicular motions) is not area-covering. The resulting gaps in each sweep area led to an effective sky coverage of 66%. This explains too, why Herschel has missed some objects. Due to catalogue identities, he actually discovered 2,367 deep-sky objects (with a brightness of 12.1 mag). However, the sky offers 3,585 objects, accessible with the 18.7-inch reflector. Thus 1,218 objects were missed, which gives a success rate of 66%. This exactly corresponds to the observed sky, resulting from his not area-covering sweep method. Thus, Herschel worked with maximum effectivity! Moreover, given the rough mounting, the data quality and position accuracy are astonishingly high.

The catalogued objects are sorted by their visual appearance (star cluster, nebula) and brightness in 8 classes (I–VIII); within a class, they are numbered by discovery date. For instance, class IV ('planetary nebulae') has 78 entries, starting with IV 1 = NGC 7009 in Aquarius. For Herschel, it was a repository for objects, not fitting into any other class. Obviously, the visual appearance of Jupiter and Uranus brought him to call similar looking nebulae 'planetary', i.e. objects with an even light, forming a disc with definite edge.

In Herschel's catalogues, galaxies are clearly dominating (much more than in the Messier catalogue). We have 89% galaxies, 7% open clusters; the rest are diffuse/planetary nebulae and globular clusters. The brightest object is NGC 2264 (4.1 mag), an open cluster in Monoceros; the faintest NGC 2843 (15.5 mag), a galaxy in Cancer. The latter again demonstrates the power of the reflector and, of course, Herschel's outstanding observational skill.

The 15.5 mag galaxy NGC 2843 in Cancer is the faintest object in Herschel's catalogue. It was found in 1787.

Between 1785 and 1789, Herschel, now living at Slough, constructed a 'front-view' reflector of 48 inches aperture and 40 feet focal length. This enormous task absorbed much of his time, needed for further sweeps. It was the largest telescope in the world for about 50 years (beaten in 1845 by Lord Rosse's 72-inch reflector). Due to its cumbersome structure and unexpected optical defects, it was neither used for sweeping nor the study of nebulae. However, Enceladus and Mimas, the 6th and 7th moon of Saturn, were discovered in 1789 with the '40 foot'. It was soon clear that the huge instrument could not fulfil the high expectations of its maker (and the financier, King George III). In 1840 it was eventually dismantled by John Herschel.

John continued the observations of his father in two campains. The first covers the period 1825–33 at Slough. He used a front-view reflector with an aperture of 18¼ inches and 20 feet focal length, built in the style of William and completed in 1820. John's primary intention was not finding new objects but the re-examination of the three Herschel catalogues. All discoveries should be confirmed, giving better positions and descriptions. To realise this project, he compiled 'working lists' to direct his sweeps, making the observations more effective. They are based on Caroline's unpublished *Zone Catalogue*. However, during 428 sweeps, many new nebulae and clusters were found. The observational result (Slough catalogue) is published in the *Philosophical Transactions of the Royal Society* for 1833.

The second campaign was a systematic survey of the southern sky, made 1834–38 at Feldhausen, South Africa (now a part of Cape Town). The southern mission, again using the 18¼-inch, was much different from the Slough campaign, inspecting mainly terra incognita between −2.5° and −90° declination. Only Halley, Lacaille and James Dunlop (Australia) had already observed parts of the southern sky, though with much smaller instruments and not in a comprehensive manner.

John Herschel's 18¼-inch reflector, erected at Feldhausen, near the Table Mountain.

At the rural site, John performed 382 sweeps in 349 nights. The result, known as the Cape catalogue, was not published until 1847. It contains 1,733 entries. The objects are mainly galaxies (56%) and open clusters (30%). Counting the independent deep-sky objects we get 1,649, of which 1,125 were new.

There are two reasons for the discovery of the large number of open clusters, firstly the Milky Way is much more prominent in the southern sky and, secondly, we have the Magellanic Clouds, offering a bunch of individual nebulae and clusters. As 638 possible deep-sky objects that he could have seen were missed, John Herschel's success rate is 72%. As in the case of his father, the sweeping method does not lead to a complete coverage. The mean visual magnitude is 12.2 mag. John also identified clusters of galaxies. For instance, he recognized the crowded regions in Centaurus and Fornax. What makes the younger Herschel unique is the fact that he is the only observer in history who visually surveyed the entire sky with a large telescope. In 1864, he eventually published a catalogue of all known nebulae and star clusters, the *General Catalogue* (GC), listing 5,079 objects.

John Herschel's GC paved the way for the most important catalogue of non-stellar deep-sky objects, the *New General Catalogue* (NGC), published in 1888 by the Danish astronomer John Louis Emil Dreyer. The Danish astronomer had mainly worked in Ireland and England. For some years he was scientific assistant of Lawrence Parsons at Birr Castle before he became Director of the Armagh Observatory. Dreyer also is the author of the monumental *Scientific Papers of Sir William Herschel*, published 1912 in Oxford.

The Danish astronomer John Louis Emil Dreyer (1852–1926) is the creator of the famous *New General Catalogue* (NGC), first published in 1888.

With its 7,840 entries, the NGC is the last comprehensive visual catalogue of non-stellar objects covering the whole sky. Despite all modern successors, which usually feature a particular type of object (galaxies, open clusters etc.), it is still the most popular catalogue and the NGC number is the primary designation for bright, large deep-sky objects, both in amateur and professional astronomy. In 1895 and 1908, Dreyer published two supplements (IC I and II) with a total of 5,386 objects, today collectively referred to as the *Index Catalogue* (IC). The IC II is entirely based on photographs. The Dreyer catalogues are abbreviated as NGC/IC.

First Ideas about Star Clusters and Nebulae

Pierre-Simone de Laplace and Immanuel Kant had thought about the formation of cosmic bodies by gravity, based on Newton's revolutionary theory of 1687, explaining all mechanical processes on Earth and beyond. Laplace and Kant theorised, the solar system should have formed out of spinning nebulous matter, which became oblate by the centrifugal force and eventually condensed into Sun and planets. Saturn, showing a central body surrounded by a ring, was seen as a proof – an object in the making.

However, not only was the question of evolution of heavenly bodies tackled. Another challenge concerns the distance and hierarchy of cosmic objects like the Andromeda Nebula. The first important contribution to the issue was made by Kant, who interpreted these objects as independent 'Milky Ways'. In 1754, he wrote: 'We have seen with astonishment figures in the heavens, which are nothing but those systems of fixed stars, limited to a common plan, such Milky Ways, if I may express myself in this way, which represent elliptical forms, seen by the eye in various positions and appearing faint due to its infinite distances.'

William Herschel's opinion on this point is controversial. There is no definite statement from him that 'extragalactic nebulae' existed. On the question of the nature and evolution of cosmic objects, his view was much clearer. He was well aware of the Kant-Laplace theory about the origin of the solar system and early thought about its conclusions for the starry heavens. For Herschel, it was also likely that stars and clusters should have formed from nebulous matter by a process of condensation and rotation due to gravitational action at a distance. He was convinced that powerful telescopes would bring the required evidence, i.e. the visual detection of change and evolution in the realm of nebulae and star clusters. Perhaps not for a single object – because of the supposed slowness of cosmic processes – but for a collective of objects, found in different stages of evolution. Thus, it becomes a matter of correct interpreting the recognized forms – strongly in the tradition of 'natural history'.

Herschel's early observations of the Orion Nebula ('the most beautiful object in the heavens') with his 6.2-inch reflector in Bath already convinced him that true nebulous matter exists ('nebulosity of the milky kind'). This was based on supposed shape and brightness changes of its parts when comparing his observations with earlier ones. Further support of this idea was brought by the resolution of some 'round nebulae', like the globular cluster M 30 in Capricornus. Using a 12-inch reflector, he wrote in August 1783: 'Plainly resolved into very small stars. It is a difficult step, i.e. if we divide the transition from the Pleiades down to the Nebula of the Orion into six steps this is perhaps the 4th towards the real nebulae.' Obviously, the evolution was marked by objects in different stages (much like living things).

Curiously, based on observations of the Omega Nebula (M 17) and Dumbbell Nebula (M 27) with the 18.7-inch, Herschel later changed his point of view. For M 17 he wrote in 1784: 'the milky nebulosity seems to degenerate into the resolvable kind [...] this nebula is a stupendous Stratum of immensely distant fixed stars'. M 27 was described as 'double stratum of stars of a very great extent [...] The ends next to us are not only resolvable nebulosity but I really do see very many of the stars mixt with the resolvable nebulosity.' Herschel now arrived at the conclusion that all nebulae must be star clusters – and their resolution would only be an issue of distance and aperture. In 1785, he developed a new evolutionary scenario: the universe started with widely distributed stars, which slowly condensed to larger agglomerations ('stratum', 'Milky Way') by gravitational forces, eventually fragmenting into many smaller clusters. The density reaches its highest degree in globular clusters, ending as planetary nebulae, which 'may be looked upon as very aged [globular clusters] drawing on towards a period of change, or dissolution'.

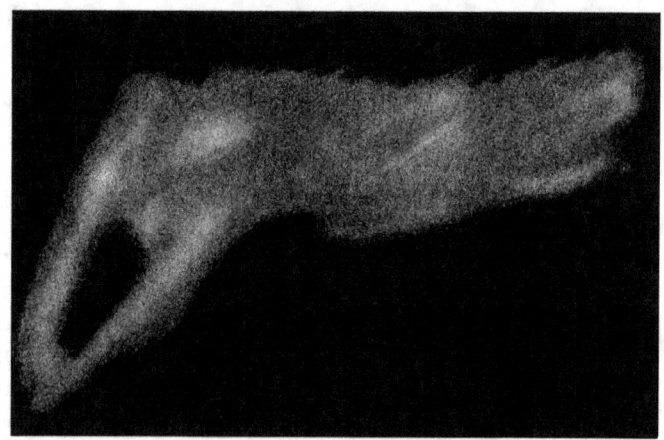

William Herschel's illustration of NGC 6992, the western part of the large supernova remnant in Cygnus, known as Veil Nebula.

However, an observation made on 13 November 1790 led Herschel to change his view again. He discovered the planetary nebula NGC 1514 in Taurus: 'A most singular phenomenon! A star of about 8th magnitude with a faint luminous atmosphere.' What follows was a revision of his current idea that all nebulae should be clusters. The dominant central star seems to be strongly correlated with the surrounding nebula and must therefore be formed by gravitational contraction. As explained in his paper *On nebulous stars properly so called*, Herschel now was convinced that at least some of the unresolved nebulae consist of a 'self-luminous fluid'. According to this (final) hypothesis, true nebulosity would gradually condense into stars or clusters. The cosmic version of the Kant-Laplace theory, supported by Herschel and others, was later called the 'nebular hypothesis'.

 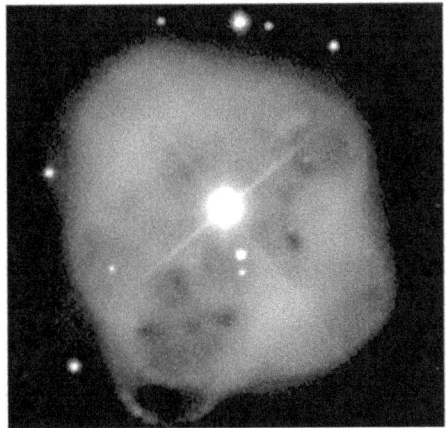

The bright planetary nebula NGC 1514 in Taurus. Left: Herschel's sketch of his 'star with an atmosphere' (1790); right: modern image (the star shines at 9.5 mag).

Though inspired by his father, John Herschel's cosmology was not an evident theory. He often changed his arguments and ideas. Once more, the basic question was the physical relation between nebulae and star clusters, focused on the terms 'resolvability' and 'change'. John claimed to have attacked the issue by observational facts, whereas his father, in his opinion, rested on speculation. He was cautious, for the nebular hypothesis touched a fundamental conflict between religion and science – a delicate matter. The theological doctrine still declared a static system of the world, in which God created the stars in eternal spheres. In this picture, there was no room for 'materialistic' ideas involving nebulous matter and cosmic evolution.

Lord Rosse's Spiral Nebulae

In the nineteenth century, Birr Castle, located in the heart of Ireland, was among the most important centres for visual observations of nebulae and star clusters. Its large reflectors of 36 inches and 72 inches aperture were used for systematic research from 1839 to 1878. Their constructor, William Parsons (Lord Rosse, since the death of his father in 1841), was multi-talented, like William Herschel. As an engineer of great skill, the Irish noble man naturally was inspired by Herschel's large reflectors. Because this eminent ethnic German astronomer had not published details about their making, Parsons had to reinvent most of the complex procedures: from casting the metal mirror to the telescope's optical and mechanical design. The heavy mirrors (known as 'speculum') were made of an alloy of copper and tin, giving a reflectivity of up to 70%.

William Parsons (1800–1867), the 4th Earl of Rosse, known as Lord Rosse. The eminent engineer and astronomer erected the Birr Castle observatory, right in the centre of Ireland.

In September 1839, the 36-inch reflector ('3-foot') was erected, having almost twice the aperture of Herschel's standard instrument. Then – by doubling the aperture – engineering was pushed to the limit in the shape of the celebrated 'Leviathan of Parsonstown'. The 72-inch reflector ('6-foot') with 54 feet focal length was operational in spring 1845. It was the world's largest telescope for many decades (a larger one was not operational until the 100-inch Hooker reflector on Mount Wilson in 1917). Visually, the optical power is comparable to a modern 25-inch reflector, equipped with an aluminised glass mirror.

Like Herschel's observatory in Slough, Birr Castle was a private institution. Its wealthy owner was free to use the giant instruments. Lord Rosse followed the pioneering work of the Herschels, whose field of activity was not the classical astronomy (measuring positions of comets, planets or stars, routinely done by professional observatories), but some kind of 'astrophysics'. His aim was to view and draw bright objects with the largest possible aperture. The ideal data source for this task was John Herschel's Slough catalogue of 1833, containing drawings of interesting objects. About 1840, Parsons started a comprehensive, high-resolution study with the 36-inch, later continued with the 72-inch. As a by-product, many new objects were discovered in the vicinity of the targets. He interpreted the results with scientific strength – again following his idol, William Herschel.

Birr Castle observatory about 1845. Infront, between the massive walls, is the famous 'Leviathan of Parsonstown', the 72-inch reflector; behind its forerunner, the 36-inch.

In the first years, two friends and experienced astronomers often visited Birr Castle to use the large reflectors: The Reverend Thomas Romney Robinson, Director of Armagh Observatory, and James South, an experienced double-star observer and owner of a private observatory in Kensington. From 1848 until his death in 1867, Lord Rosse employed scientific assistants – a response to the pressure caused by his various official and scientific duties (for some years he even was President of the Royal Society). The assistants made most of the observations, resulting in careful descriptions and drawings. The results appeared in the *Philosophical Transactions of the Royal Society*. In 1880, a review of all Birr Castle observations was complied by Dreyer, who worked as scientific assistant of Rosse's son and successor Laurence Parsons.

What was Lord Rosse's scientific motivation to build large reflectors? He simply wanted to answer the fundamental question: are all nebulae resolvable? This would eventually solve the conflict about the nebular hypothesis. The major protagonists were the Irishman Robinson and the Scot John Pringle Nichol. Nichol, Director of Glasgow Observatoy, who was a glowing pendant of the hypothesis. He is also responsible for the term 'galaxy' in the sense of an independent stellar system outside the Milky Way. Under the impression of the discoveries in Birr Castle, Nichol described 'remote galaxies', which, however, were initially concerned with globular clusters.

His eloquent opponent, Robinson, was convinced that all arguments for Herschel's 'self-luminous fluid' could be destroyed by observation. Thus, the Reverend was eager to get the desired result: the resolution of nebulae. Thanks to Lord Rosse's immense skill in constructing large telescopes, Robinson had powerful weapons in his fight against Nichol. This conflict mainly influenced the focus of research at Birr Castle. Alas, his chronic aversion for the nebular hypothesis sometimes led him to wishful thinking – against any empirical evidence. Lord Rosse often had to calm his friend. Though the Protestant Rosse was a member of the Church of Ireland, he always acted as an independent, free-thinking scientist. His goal was to prove whether a large reflector could reveal a starry nature for such intractable targets as the Orion Nebula. In case of failure, he would not hesitate to accept true nebulosity. It actually did.

Rosse's drawing of M 51 in Canes Venatici, known as the 'Whirlpool Nebula' (1845).

A crucial point in this story was Lord Rosse's discovery of the spiral structure of M 51 in early April 1845, using the 72-inch. It is a mysterious fact that

Robinson and South failed to see this structure a few weeks before with the brand-new instrument. Lord Rosse, observing alone, made a sketch, which he presented shortly after at a scientific meeting held in Cambridge, chaired by John Herschel. Of course, the detection of spiral structure would affect the nebular hypothesis. A spiralling mass fits surprisingly well to Herschel's idea of a spinning nebula from which stars and clusters has formed, so strongly supported by Nichol. Thus, Lord Rosse's spiral nebulae were even better evidence for true nebulosity than was the Orion Nebula against it – evolution matters more than resolution. However, Robinson tried to rescue his ideology by a clever reinterpretation of the observations. Denying the existence of nebulous matter, he simply postulated a rotating ensemble of stars: 'Their resemblance to bodies floating on a whirlpool is, of course, likely to set imagination at work, though the conditions of such a state are impossible there. A still more tempting hypothesis might rise from considering orbit motion in a resisting medium; but all such guesses are but blind.' This is the origin of the name 'Whirlpool' for M 51.

In the end, all could live with the new situation. Forever the diplomat Rosse did not speculate about the nature of the spiralling mass but stated that the observed pattern 'indicates the presence of dynamical laws'. He further wrote 'that such a system should exist, without internal movement, seems to be in the highest degree improbable'. From the modern point of view, there are true elements in both ideas. The spiral arms host stars as well as 'nebulous matter' (gas, dust). However, according to the density-wave theory, the arms themselves are not truly rotating (they behave like a Mexican wave). Furthermore, the matter does not wind up in the centre, as in a whirlpool, and no stars are born in the hub (their cradle is in the spiral arms).

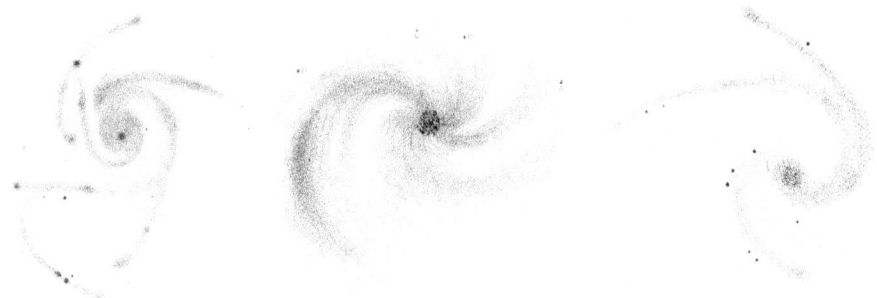

Drawings of spiral nebulae, made at Birr Castle: M 101, M 94, and M 33.

In the following period, Lord Rosse and his assistents searched for other 'whirlpools', finding spiral structure in 76 objects – an astonishing number of which (67) are actually spiral galaxies. Laurence Parsons continued his father's research, inspecting and discovering many nebulae together with his talented assistants John Dreyer and Ralph Copeland.

A diversity of physical structures in galaxies was seen with the 72-inch for the first time: spiral arms, bars, absorption bands or knots (star forming regions). Although the search for new nebulae was not on the Birr Castle agenda, 330 objects were eventually found over a period of 30 years, mostly in the vicinity of other objects. Dreyer later entered them in the NGC/IC. How much the 72-inch has expanded the visible universe is shown by the faintest objects – invisible in all other telescopes of that time. If 12.2 mag was the mean brightness of the Herschel objects, the 'Leviathan' pushed it to 14.3 mag; the faintest object even reaches 16.4 mag (the galaxy NGC 2689 in Ursa Major, found 1858).

Beside the great achievements, like the discovery of spiral structure, there were failures too. One concerns the 'resolvability' of nebulae, influenced by ideological constraints. Nowadays it sounds rather curious that gaseous nebulae like M 42 or galaxies like M 51 and M 31 should have been considered 'resolvable'. The Birr Castle observers entertained an illusion. Even the 72-inch was unable to resolve galaxies into single stars – they are simply too remote.

No doubt, the subjectivity of visual observing and drawing is the main reason for the controversial discussions among the nineteenth century astronomers. Some even denied the existence of spiral nebulae – simply because their inferior instruments did not show them. Of course, there were no reliable images, and the physical nature of the nebulae was still unknown. Lord Rosse was always aware of these problems and acted in a scientific manner – others did not!

Spectroscopy, Photography and the True Nature of Galaxies

The endless discussions among the nineteenth century astronomers about the physical nature and evolution of nebulae and star clusters led to the frustrating cognition that this important issue could never be tackled by visual observations. Neither the large telescopes, with their enormous light-gathering power, nor extensive measurement campaigns to determine precise positions were able to terminate the speculations and controversies. Fortunately, the scientific breakthrough was soon to come: the new field of 'astrophysics', namely spectroscopy and photography.

It was already known that the Sun and stars show a continuous spectrum with dark (absorption) lines. The Sun was first studied by Joseph Fraunhofer in 1814 with a simple prismatic device. On the other hand, there exist line spectra which are strongly correlated with chemical elements. The distribution and relative strength of these lines are different for each atom in the periodic system.

The pioneering work on spectral analysis was done in 1859 by Robert Kirchhof and Robert Bunsen.

In 1864 the revolutionary method was introduced to deep-sky astronomy by the English amateur William Huggins. With a spectroscope, attached to his 8-inch refractor, he could visually inspect the spectra of bright nebulae. In the case of the Andromeda Nebula he detected a continuous spectrum with some absorption lines, thus the object actually consists of stars. Due to its extend and brightness it must contain large numbers, very far away. In 1889 the American astronomer Charles Young described objects like M 31 as 'white nebulae' (the resulting colour of a continuous spectrum is nearly white) and in 1911 it was asked: 'are white nebulae galaxies?'

The spectrum of the Orion Nebula appeared much different in Huggins' instrument, showing single emission lines instead of a continuous spectrum. They are due to hydrogen. Thus, M 42 contains hot gas – true nebulous matter in the sence of the nebular hypothesis exists! We now know that stars were formed from it. However, the process of contraction is far too slow to be visually observable. Thus, Herschel's claimed 'change' in the shape of the Orion Nebula was pure illusion. Huggins also inspected several planetary nebulae, for instance, the Cat's Eye Nebula (NGC 6543) in Draco shows lines of oxygen and nitrogen.

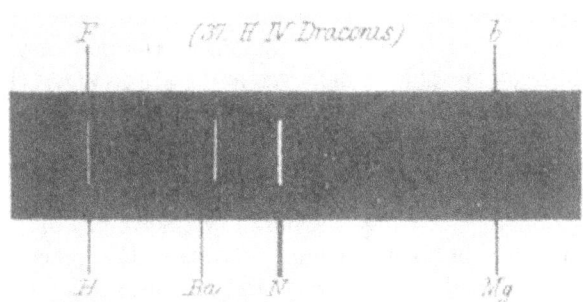

Spectrum of the planetary nebula NGC 6543 in Draco, drawn from visual inspection by William Huggins. It shows lines of hydrogen and nitrogen.

Spectroscopy was soon applied by a number of progressive astronomers, while traditionalists could not accept its value. It is interesting that William Pickering and Ralph Copeland reversed the method. Both tried to unmask a 'star' as a planetary nebula by its line spectrum, seen in the spectroscope. They visually detected 22 starlike planetaries (included in the NGC).

In 1840 the new technique of photography was applied to astronomy. Due to the insensitive plates, only the brightest objects were chosen. The American astronomer Henry Draper started with an image of the Moon; the French physicist Léon Foucault followed with the Sun. The first picture of a star was made in 1851 by the American pioneer of photography John Adams Whipple, it shows Vega. Due to their faintness, nebulae had to wait until 1880, when

Draper could portray the Orion Nebula. Lord Rosse's Whirlpool Nebula M 51 was photographed nine years later by the English amateur Isaac Roberts. The long exposure times were a real challenge (often several nights were needed). Not only refractors (astrographs) or reflectors were used, but also small portrait lenses, giving a large field. Photography was later combined with spectroscopy, recording the spectrum on a sensitive plate. Another astrophysical method, the photometry of nebulae, was a difficult task. Reliable brightness values, measured by electric currents, could be determined not until the twentieth century.

The first image of M 51 was taken by the English amateur Isaac Roberts in 1889 (north to the right).

The 1920s were the time to finally clear the nature and cosmic origin of spiral nebulae. The issue culminated in the 'great debate' of 1924, held between the eminent American astronomers Harlow Shapley and Heber Curtis. For Shapley, the objects belonged to the Milky Way, which he regarded as a gigantic system ('big galaxy'), which he even identified with the entire universe. For Curtis, however, the spiral nebulae were 'island universes' and he spoke deliberately of galaxies.

Shortly thereafter, the puzzle was solved by the American astronomer Edwin Hubble – in favor of Curtis. Starting in 1925, the Hubble astronomer could detect Cepheid variables in several nebulae, using the 100-inch reflector on Mount Wilson. By measuring their light curves, Hubble could apply the celebrated period-luminosity relation to determine the distance. In the case of M 31 he got an immense value of 800,000 light-years – the Andromeda Nebula is far outside the Milky Way. However, Hubble's results were later significantly corrected. Actually, M 31 is 2.5 million light-years away.

More than a decade before the first distance to a galaxy was measured, an unexpected result had already been provided for headlines. In 1912, Vesto Slipher, an astronomer at the Lowell Observatory in Arizona, started a campaign to measure the spectra of spiral nebulae. It is based on the research of Christian Doppler, done in the 1850s. The Austrian physicist had revealed that the wavelength of light (correlated with its colour) depends on the relative velocity between source and observer. The higher the velocity, the larger the wavelength-shift. The Doppler Effect appears as a shift of all spectral lines towards the blue or red end of the continuous spectrum. Its amount gives the radial velocity, its sign the motion direction: + towards the observer (blueshift), – away from the observer (redshift). Until 1915, Slipher had obtained the spectra of 14 galaxies, yielding a curious result: all but one showed a redshift. Interpreted as a Doppler Effect, 13 objects are radially moving away from us with velocities up to 1,100 km/s (the exception is M 31 with –300 km/s). In 1924 already 24 cases were known, confirming the previous result.

Meanwhile, Hubble had determined distance values for the galaxies of Sliphers' sample – and ingeniously combined them with the measured redshifts. Astonishingly, there was a linear relationship: the greater the distance, the larger the redshift. This is the famous Hubble-Lemaître law, eventually published in 1929. The proportionality factor is called 'Hubble parameter'. With the Doppler Effect the interpretation sounds simple: galaxies have a radial velocity which increases with distance. In the 1930s, Hubble and his assistant Milton Humason discovered a galaxy showing the enormous 'recession velocity' of 61,000 km/s. But what causes this strange motion? Here Albert Einstein comes into play (the issue will be cleared in a later section).

Hubble and Humason measured the redshifts of many nebulae to confirm the 'velocity-distance relation' (Hubble-Lemaître law). The estimated distance results from the appearent magnitude of the dominating galaxy in a cluster.

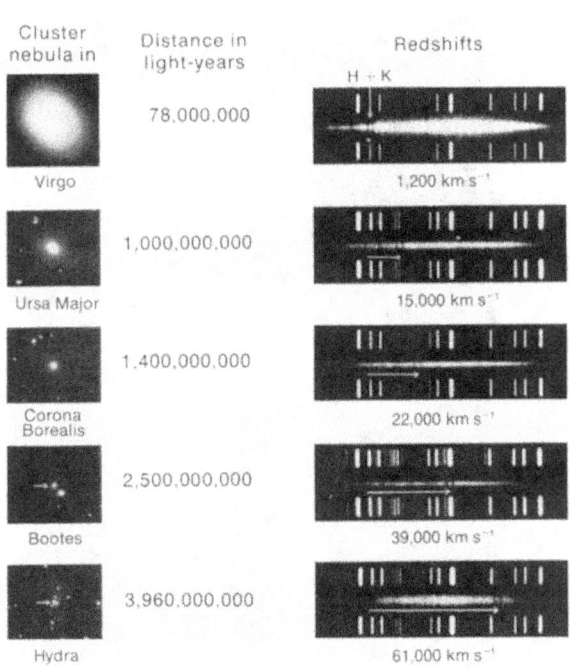

Reversing the Hubble-Lemaître law, we have an ingenious method to determine the distance of a galaxy, reaching far into the cosmos. It is therefore possible to draw a three-dimensional map, showing the galaxy distribution. Smaller groups were found, like the Local Group (including the Milky Way, M 31 and M 33), and large aggregations with thousands of galaxies, like the Virgo Cluster. There are even clusters of clusters of galaxies, called superclusters. This led to the idea of cosmic hierarchy. The universe is built of stars, arranged in galaxies, which are members of galaxy clusters, forming – in the last step – superclusters. Between these entities, there is empty space. The matter in the observable universe is distributed like an irregular network with massive knots (galaxy clusters), long filaments (formed by galaxy groups) connecting them, and voids. The large-scale structure of the universe looks much like foam or Swiss cheese.

Thanks to the work of astrophysicists and constructors of large telescopes – and even amateur astronomers – we now know much about the content, structure and evolution of the universe and the issue is no longer subject to speculation. Still leaving our home not farther than the Moon, we were able to push our knowledge to the remotest places. This is due to the universality of the physical laws and the information brought to use by various kinds of radiation and, finally, to the (limited) speed of light, which let us look far into the past.

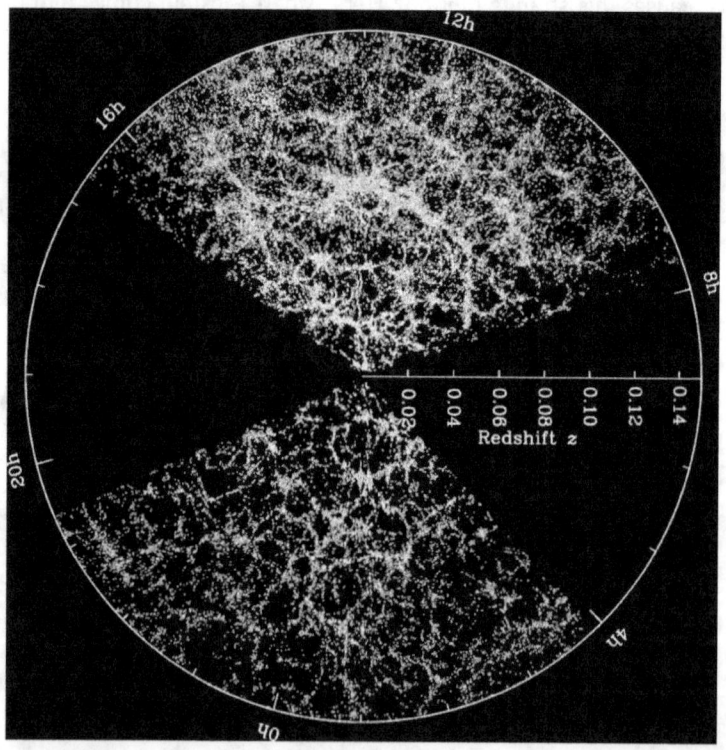

The large-scale distribution of matter (galaxies) up to distances of 600 million light-years, based on redshifts, measured in the Sloan Digital Sky Survey (SDSS).

Object Types and Their Astrophysical Nature

General Data of Deep-sky Objects

The following table shows the types of non-stellar deep-sky objects, presented in this book. Stars (single, double, multiple or variable) are not treated. The primary distinction is between galactic and extragalactic objects, i.e. belonging to the Milky Way or beyond. There are many cases, showing a combination of types, e.g. open cluster/diffuse nebula or emission/reflection nebula. Concerning planetary nebulae, the type definition is historically motivated; they are emission-line objects. Not all types are strictly non-stellar; a typical exception is the quasar.

Types of deep-sky objects (with abbreviations, used in this book)

Location	Type	Subtype	Special objects
galactic	star cluster	open cluster (OC)	association, moving group
		star group	pattern (asterism)
		globular cluster (GC)	
	diffuse nebula	emission nebula (EN)	HII region, supernova remnant (SNR)
		reflection nebula (RN)	bipolar/cometary nebula (BN)
		dark nebula (DN)	HI region
		planetary nebula (PN)	annular nebula
extragalactic	galaxy	ordinary galaxy (GX)	spiral, elliptical, irregular, peculiar
		quasar (QSO)	BL Lacertae object
		object in galaxy	HII region, super star cluster, globular cluster
	multiple galaxy	pair	optical/physical
		group	trio, chain
	galaxy cluster	rich galaxy cluster	elongated, spherical
		supercluster	

The following sections focus on information, useful to localize an object (position, orientiation) or to quantify its apparent features (brightness, size). Further, data sources and catalogues of deep-sky objects will be presented. Concerning the latter, one has to distinguish between general compilations (like the NGC/IC) and object-specific catalogues, which focus on a special type (open cluster, galaxy etc.).

Celestial Position and Orientation

The celestial position is usually given by coordinates. Most common are right ascension (RA, α) and declination (Dec, δ). Since the underlying reference system is based on the celestial equator, we speak of equatorial coordinates. An equatorially mounted telescope has a RA- and Dec-axis.

Right ascension is given in hour, minute, and second of time (^{h} ^{m} ^{s}), running from 0^h to 24^h (west to east); note that east is left in the sky (in contrast to the Earth atlas). The zero point is at the intersection of celestial equator and ecliptic, the vernal equinox (currently in Pisces). Declination is given in degree, arc minute, and arc second (° ' "), running from –90° (south celestial pole) over 0° (celestial equator) to +90° (north celestial pole). At the celestial equator, a minute of RA equals 15 arc minutes, so the scales of α and δ are not equal. Towards the poles, the relation becomes smaller; for instance, at $\delta = 80°$ we have $1^m = 2.5'$.

Other reference systems are in use; we have horizontal, ecliptical, or galactic coordinates. All systems can be transformed into each other by simple formulae. The horizontal system with azimuth and altitude refers to the Earth. The altitude (elevation) is the angle above the horizon, running from 0° to 90° (zenith). The azimuth is the horizontal direction, running from 0° (south), 90° (west), 180° (north), to 270° (east). Most Dobsonian reflectors and some Schmidt-Cassegrain telescopes (SCT) have an altazimuthal mounting.

The galactic system is used to localize objects in the Milky Way. The galactic length runs along the Milky Way plane (zero point = galactic centre in Sagittarius), the galactic latitude is perpendicular to it (+90°/–90° = north/south pole of the Milky Way, located in Coma Berenices/Sculptor). Galactic coordinates are also usefull to determine the position of extragalactic objects relative to the Milky Way. This may show the spatial distribution of nearby galaxies, e.g. of Local Group members.

The equatorial frame is not constant in time, because the direction of the Earth's axis is slowly moving in a circle around the pole of the ecliptic, known as precession (the period is about 26,000 years). This motion both shifts the celestial equator and the vernal equinox. Thus, the values of RA and Dec slowly change with time. Consequently, for the coordinates, the time of observation (epoch) should be added. To avoid this complication, the equatorial system usually refers to a standard time (equinox). The current choice is the beginning of the year 2000. In older catalogues or sky atlases we find positions for 1950 or even 1900. One can convert (precess) the coordinates for different equinoxes. This is an easy task for a planetarium program. Of course, objects can actively change the position by their space motion. Thus, in precise star catalogues, the proper motion must be given.

If a place should only be roughly fixed, coordinates are not necessary. A relative position to the neighborhood star will do. Here the constellation comes into play. Brighter stars are localized by Bayer letter or Flamsteed number plus constellation symbol, like ζ UMa (Mizar in Ursa Major). More quantitative are position angle PA (in °) and angular distance (in '). PA is the angle between the line from the main object to the target and the north direction; it runs from 0° (north), 90° (west), 180° (south) to 270° (east). Take, for instance, as basis the bright star β And (Mirach) and view at PA 328° (northwest) in a distance of 6.8' – you will find the small round galaxy NGC 404. The professional tool to measure relative positions is the micrometer. For an estimate, an ordinary eyepiece can already do. Two things must be known: the diameter of the field of view and its orientation. The latter not only depends on the celestial position, but also on the telescope: the field can be upside down (refractor) or even mirror reversed (reflector). Things are easier for an eyepiece with crosshairs.

The small round galaxy NGC 404 near the bright star β And is known as Mirach's Ghost.

The position of a pointlike (stellar) object is clearly defined. But what about an extended one? In case of symmetry (round or elliptic shape) there is a definite centre. Examples are regular open clusters, globular clusters, elliptical and lenticular galaxies or even many spirals. The precision of the coordinates relates on the definition of the centre. The worst cases are asymmetric or irregular systems, like NGC 55 in Sculptor (asymmetric) or NGC 4449 in Canes Venatici (amorphous). Precise coordinates are meaningless now.

Brightness, Magnitude, and Size

Brightness is one of the most relevant parameters, for any kind of observation. Basically, one must distinguish between apparent and absolute brightness. The apparent brightness of an object (designated m), is that seen by an observer on Earth. It is measured in magnitudes (mag). Note that m = 9 mag is brighter than m = 10 mag. Consequently, zero and negative magnitudes are used for very bright objects (Venus reaches m = –4.4 mag). Brightness is proportional to the logarithm of the intensity. The proportionality factor has been set to 2.5 (a tenfold intensity means a magnitude increase of 2.5). This corresponds to the logarithmic sensitvity of the eye.

Absolute brightness (M) is easily defined – but difficult to determine. The important quantity is needed, because all celectial objects have different luminosities; stars, for instance, are far from being equal bright sources. The idea: put, hypothetically, every object at a standard distance of 10 parsecs (32.6 light-years). Then it is easy to see which is truely brighter or fainter. Of course, to determine the absolute brightness, the objects distance must be known. Absolute brightness (measured in mag) refers to the physical state of the object; it is equivalent to luminosity (measured in Watt).

For apparent brightness we must distinguish between integrated and surface brightness, based on two types of sources: point-like and extended. For a point source, like a star or quasar, the light appears naturally integrated due to the great distance. But an integrated brightness can also be assigned for an extended source like a galaxy. Now the light is focused to a point by a lens and then compared with a reference star of known (integrated) brightness. When talking about the apparent brightness of a celestial object, it usually means the integrated quantity. With the surface brightness, you go the opposite way. To quantify the brightness of an area, a reference star is defocused onto a unit area. The surface brightness is then defined as the apparent brightness per area (measured in mag per square arc minute).

This requires the definition of apparent size. For round objects, like a planetary nebula or globular cluster, only a single value is needed, the diameter (a), usually measured in arc minutes (') or arc seconds ("). For elongated objects, like elliptical or spiral galaxies, the small and large diameter is given. The apparent size is written as a × b. Of course, this measure does not work for irregular objects. In this case one takes the maximum ellipse enclosing it and measures the axes.

Brightness is usually related to a specific spectral range (colour); when collected over all wavelengths, one speaks of bolometric brightness. Most commonly used is the UBV system, defining the brightness in the standard colours ultraviolet, blue and yellow. The yellow value is also called visual magnitude (V or m_v), because the eye has its maximum sensitivity at that colour. It is the standard for visual observers. To decide if an object can be seen, you need the V magnitude. A blue magnitude (B or m_B), is often given in object-specific catalogues. For it can be determined from an image, it is also termed photographic magnitude (m_{pg}). This value says nothing about observability. For instance, galaxies can visually be about 1 mag brighter than on a photograph. That means, you have to check, which spectral magnitude is listed; there are also red (R) or even infrared (I) values.

Data Sources and Catalogues

The main sources for deep-sky objects are the Messier catalogue and the NGC/IC, covering all types. In the 1990s it was discussed, if the Messier catalogue with its 110 entries should be further extended. Indeed, there are objects, which – by their brightness – had the right to be included. Examples are the open cluster NGC 188 in Ursa Minor and the planetary nebula NGC 40 in Cepheus (both are very near to the northern celestial pole). The existence of such objects inspired the popular English amateur Patrick Moore to publish a list of the best non-Messier objects in 1995. The result, called *Caldwell Catalogue*, has 109 entries (C 1 to C 109), covering the entire sky. Curiously, it is sorted by decreasing declination, thus we have C 1 = NGC 188, C 2 = NGC 40; the final object, C 109 = NGC 3195, is a bright planetray nebula in Chamaeleon. Moore's compilation, however, never reached the popularity of the Messier catalogue.

All NGC objects, but one, were found visually (22 of them with the aid of a spectroscope). The exception is NGC 1432, the faint nebula around Maia in the Pleiades, detected on a plate by the Henry brothers (Paris Observatory) in 1885. The IC is mainly a product of photography. On average objects in the NGC are 0.5 mag brighter than those in the IC I and 1.5 mag brighter than those in the IC II. The latter, incidentally, contains two Messier objects: the star cloud M 24 = IC 4715 and the star cluster M 25 = IC 4725 (both located in Sagittarius).

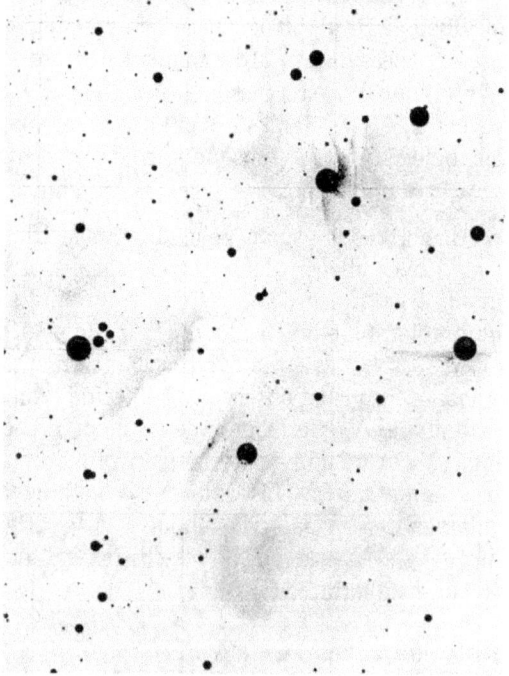

The first NGC object found by astrophotography: the Maia Nebula (NGC 1432) in the Pleiades. Below is the Merope Nebula (NGC 1435), found visually by Wilhelm Tempel in 1859.

However, the original form of the NGC/IC is not very userfriendly. The object description follows a standard already introduced by Herschel. In the case of Dreyer's *New General Catalogue*, the 7,840 'objects' were discovered by 100 observers with different instruments at various locations. Thus, it is a quite inhomogeneous work, which has led to problems. 263 entries must be deleted because of identities between different NGC numbers. 88 could not be verified (among them the mysterious Baxendell's Unphotographable Nebula NGC 7088 in Aquarius), 407 are stars, pairs or patterns or stars and 30 are parts of already catalogued objects. Thus, we end up with 7,052 independent non-stellar objects. The same applies for the IC: of the 5,386 entries, only 4,140 are usable.

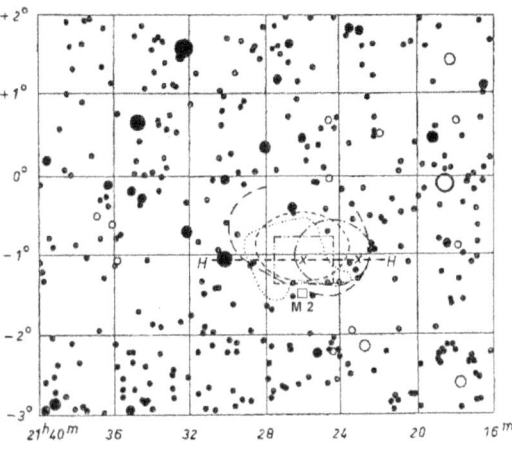

NGC 7088, a nonextistent object in Aquarius, known as Baxendell's Unphotographable Nebula.

NGC 206 is part of another NGC object: a huge star cloud in M 31 (NGC 224).

Over the years, a modern version of the NGC/IC was created by the author, the *Revised New General and Index Catalogue*. It is available on the Internet and frequent updates are made. The catalogue is largely historically correct (concerning identities and erroneous observations) and contains modern data as well as cross-references to a large number of object-specific catalogues. It also includes objects, later added to the NGC/IC and designated by an appendix (A, B etc.); an example is the polar ring galaxy NGC 4650A in Centaurus (a companion of NGC 4650).

There are modern analogues of the NGC/IC. An example is the *ESO/Uppsala Survey* of 1982 with 18,438 entries. Here, non-stellar objects south of −17.5°, seen on Schmidt camera images, are cataloged. An example of the nomenclature is ESO 29–G21, the Small Magellanic Cloud (G stands for galaxy). Moreover, the internet offers large databases. The most important of which are the *NASA Extragalactic Database* (NED) with 167 million objects (mainly galaxies) and the *SIMBAD Astronomical Database* with over 9 million (mainly Milky Way objects).

For star clusters, we have the *Catalog of Star Clusters and Associations*, compiled by the Czech astronomers Jiří Alter and Jaroslav Ruprecht. The final version lists 1039 clusters (open and globular), 70 associations, and 16 moving groups. The open/globular clusters are designated OCL/GCL plus sequence number; for instance, OCL 421 = M 45, GCL 45 = M 13. The primary reference for open clusters is the *Catalog of Open Cluster Data*, compiled by Gosta Lyngå of Lund Observatory (Sweden). It lists 1151 objects and gives information about the number of stars, diameter, type, brightness (of the cluster and its brightest member). The equivalent for globular clusters is the *Catalogue of Parameters for Milky Way Globular Clusters* by William Harris. The current version contains 157 objects.

For the dedicated observer, the book *Star Clusters* by Brent Archinal and Steven Hynes is an absolute must have: in nearly 500 pages, one finds a wealth of information and suggestions about virtually all known open star clusters, globular clusters and star patterns belonging to the Milky Way; important objects in Local Group galaxies are also added. The book gives detailed descriptions and historical notes; a special focus is on the correct identification and nomenclature of star clusters. Concerning designations, one has to mention the collection of individual lists of open/globular clusters, made by astronomers, working in the field. Important examples are Collinder (Cr, 471 entries), Melotte (Mel, 245), Stock (24), and Trümpler (Tr, 334).

For globular clusters we have the short but famous list, made by the American astronomer George Abell. The 15 objects that he and others found on the *Palomar Observatory Sky Survey* (POSS) in the 1950s bear the name Palomar. However, two of them are already in the NGC/IC: Palomar 7 = IC 1276 and Palomar 9 = NGC 6717.

The situation for diffuse nebulae is similar. A popular source is the Shapless catalogue (Sh2), puplished in 1959 by the American astronomer Stewart Sharpless. It contains 313 emission nebulae, mainly HII regions. Even larger is the *Catalog of Bright Nebulae* (LBN) by his female collegue Beverly Lynds (1965) with 1,125 objects. Indicated are type, size and data about the exciting star (brightness, spectral class). The Lagoon Nebula in Sagittarius is catalogued as M 8 = LBN 25 = Sh2-25. Other interesting lists of HII regions are due to Gum (97 objects) and Gaze & Shajn (Simeis objects, 286). The topic of emission/reflection, in connection with star formation, is treated in the works of Cederblad (CED, 215) and Herbig & Haro (Herbig-Haro objects, HH, 61).

Left: Globular cluster Palomar 9 = NGC 6717 in Sagittarius (the star is v^2 Sgr, 5 mag); right: planetary nebula Abell 81 = NGC 1454.

The primary source for supernova remnants is the catalogue of David Green (SNR). It is constantly updated; the latest version includes 295 objects. The notation is based on galactic coordinates, e.g. SNR 184.6-05.8 = M 1. In 1966 the Canadian astronomer Sidney van den Bergh listed 158 reflection nebulae (vdB). The most popular catalogue of dark nebulae is due to the American astronomer Edward Emerson Barnard. The objects are an output of his monumental *Photographic Atlas of Selected Regions of the Milky Way*, published in 1927. Barnard's catalogue contains 370 objects (B). The famous Horsehead Nebula in Orion is B 33. In 1962, Lynds summarized all known dark nebulae in her *Catalog of Dark Nebulae* (LDN), listing 1,802 objects.

Pioneering work on planetary nebulae was done by the Russian astronomer Boris Vorontsov-Velyaminov. In 1934 he compiled a catalogue of 134 objects (VV, not to be confused with his catalogue of interacting galaxies also known as VV). Later, the POSS opened up new opportunities for discovery and George Abell looked for faint, large planetaries. The resulting list contains 86 objects (Abell). Two of them are already in the IC: Abell 37 = IC 972 and Abell 81 = IC 1454.

Not all of these were planetary nebulae and some are nonexistant such as plate faults. Other observers followed, creating a plethora of different designations. An attempt to unify them was made by the Czech astronomers Luboš Perek and Luboš Kohoutek. In 1967 they colletced all known planetary nebulae (1034) in the *Catalogue of Galactic Planetary Nebulae* and introduced the designation PK, which has since become a standard. For instance, PK 63-13.1 = M 57, which means that the object is in the range between 63° and 64° galactic length and −13° and −14° galactic latitude, where it is the first entry. Currently, about 3,000 planetary nebulae are known in our galaxy. Due to their peculiar appearance, many of the objects bear names, like Ring Nebula (M 57), Dumbbell Nebula (M 27), or Blue Snowball (NGC 7662). It is worth noting that there is a second designation commonly used by professional astronomers: PN plus galactic coordinates (G); we have PN G063.1+13.9 = M 57.

The number of sources for galaxies or clusters of galaxies is large. Most modern catalogues are based on the inspection of the POSS plates. Extensive catalogues have been produced, such as the *Catalog of Galaxies and Galaxies* (CGCG) by the Swiss astronomer Fritz Zwicky and the *Morphological Catalog of Galaxies* (MCG) by Vorontsov-Velyaminov. The CGCG contains 29,378 galaxies and 9,133 galaxy clusters north of −3.5° declination. The MCG lists 31,917 galaxies north of −45° declination. In 1973 the Swedish astronomer Peter Nilson published the *Uppsala General Catalog* (UGC). It contains all galaxies north of −2.5° declination greater than 1' or brighter than m_{pg} = 14.5 mag. The catalogue has 12,940 entries. For galaxies, the UGC has established itself as the next level after the NGC/IC. A great variety of peculiar features and systems are presented in Halton Arp's *Atlas of Peculiar Galaxies* (Arp), containing 338 cases. The work is very popular among amateur astronomers.

Currently the largest galaxy catalogue is LEDA, which stands for *Lyon-Meudon Extragalactic Database*. The project began in 1989 as the *Catalog of Principal Galaxies* (PGC) with 73,197 entries, headed by Georges Paturel. The current database contains 1.5 million galaxies, presenting all available data. In 1991, the French astronomer Gérard de Vaucouleur published the *Third Reference Catalog of Bright Galaxies*. Though, with 23,022 galaxies considerably smaller than LEDA, it is still the measure of all things in terms of data quality and completeness.

The standard reference for quasars and related objects is due to the French couple Marie-Paule Véron-Cetty and Phillipe Véron: *Quasars and Active Galactic Nuclei*. The current edition presents data for 133,336 QSOs, 1374 BL Lacertae objects, and 34,231 AGN objects (active galactic nucleus). The brightness is given as B, V and R magnitudes.

In case of galaxy groups, the Hickson catalogue enjoys great popularity. The work of the Canadian astronomer Paul Hickson, published 1982, focuses on compact groups, showing 100 cases (HCG). For a wider spectrum of groups, the *Lyon Groups of Galaxies* (LGG) with its 485 entries should be consulted.

The undisputed stardard source for galaxy clusters is the catalogue by George Abell, Harold Corwin and Ronald Olowin called *The Distribution of Rich Clusters of Galaxies*, listing 4073 cases, designated as 'Abell' (not to be confused with the few Abell planetaries). In 1984, John Bahcall and Ray Soneira created a first catalogue of superclusters with 18 entries.

The compact galaxy group HCG 61 (The Box) in Coma Berenices contains NGC 4169, NGC 4173, NGC 4174, and NGC 4175 (north to the right).

Open Clusters

There are only a few objects, which appear as a star cluster to the naked eye. Among them are the Hyades and Pleiades in Taurus. They already played a role in ancient tales. For example, in some cultures the appearance of the Hyades in the early summer morning sky is associated with the onset of a rainy season. Homer spoke of the 'rainy Hyades'. For the Greek, the Pleiades were the seven daughters of the titan Atlas. The object is said to be found on the 3,600-year-old 'Himmelscheibe von Nebra'. Further interesting cases are the Coma Berenices Cluster (Mel 111) and M 7 in Scorpio, both mentioned by Ptolemy, and the o Velorum Cluster, described by the Persian astronomer Al-Sufi in the tenth century. While the first object appears as a cluster to the naked eye, the two others look nebulous.

To this category also belong Praesepe in Cancer and the Double Cluster χ Persei. It needed a telescope to determine that these objects are indeed star clusters. It was Galileo, who in 1610 recognized the true nature of Praesepe. In the following years, more star clusters were discovered, a process in which Giovanni Battista Hodierna was particularly successful.

Physical Nature

In the early twentieth century, open clusters were studied in detail, using astrophysical methods. This work is connected with the names of Melotte, Shapley, Trümpler and Collinder. Due to comprehensive surveys in the second half of the twentieth century, the number of known open clusters has increased significantly. We now know more than 1,500 objects in our galaxy, but it is believed that ten times as many are hidden behind large interstellar dust clouds, still waiting for their discovery. Infrared and radio waves, which are not absorbed, can help with this issue.

The research on the formation and evolution of open clusters influenced the theories about the birth and life of stars and the structure of the Milky Way. Stars are made of interstellar matter, located in the spiral arms. It is a mixture of hydrogen, helium, and heavier elements (in astrophysics these are generally refered to as 'metals'). In the beginning, there are huge, slowly rotating clouds. Their temperature is only 30 K (Kelvin), the mass can exceed 1,000 solar masses and the size is about 1,000 light-years. Although the density is only about 20 atoms/cm^3, such a cloud absorbs light and can be detected as a dark nebula, when lying infront of distant stars. Due to gravitational attraction, it slowly contracts. Local densities variations and the conservation of angular momentum lead to a break-up of the cloud into parts of different mass (fragmentation). The individual contraction continues and it can take 100 million years for a fragment to reach the size of the planetary system. The rotation forms a disc about the compact centre. There the concentration has generated enough heat (about 100,000 K) to produce infrared radiation.

The initial cloud has developed into a cluster of protostars. Further contraction rapidly leads to a much higher core temperature. When eventually 15 million K is reached, nuclear fusion ignites – a star is born. At this temperature, hydrogen nuclei (protons) merge to form helium. This process creates enormous energy out of mass, producing a strong radiation pressure that tames the gravitational attraction. The star is stable now.

The stars of the young cluster can illuminate the remaining matter (gas, dust) in two ways. If the stellar radiation is energetic enough, the gas is excited, emitting a characteristic light in response. We speak of an emission nebula. If the energy is too low, mainly due to a greater distance, there is only scattering of light in the dust part – we see a reflection nebula. Thus, in the early phase, we get a mix of cluster and nebula. The surrounding matter is eventually blown away by the stellar wind leaving a clean ensemble of stars.

The typical diameter of an open cluster is about 10 to 50 light-years and the number of stars ranges from a dozen to a few thousand. The average total mass is about 2000 solar masses. The open clusters belong to the population I, defined by the young stars of the galactic disc. Almost all lie near the galactic equator, only a few are more than 25° away from it. Therefore, they are good indicators of the spiral structure of our galaxy.

The lifetime of an open cluster is determined by three factors. First there is the total mass: the larger it is, the stronger the resistance to escape attempts of individual members. Another influence is due to the differential rotation of the Milky Way, inducing shear forces that can desintegrate the cluster. Finally, gravitational disturbances, due to encounters with nearby clusters or clouds of interstellar matter, can disrupt it. Typical lifetimes are in the range of 100 to 1,000 million years. More than half the open clusters do not exist for a galactic year (orbital period of the sun around the centre of the Milky Way), i.e. 220 million years. Looking at the age of the Sun (4.5 billion years) or a globular cluster (11.5 billion years), open clusters are actually youthful objects.

When investigating an open cluster, the following question is important: which star is a member and which is in the foreground or background, appearing just by chance in it? One criterion is participation in the collective motion of the cluster. Thus, one must determine the space motion of each candidate. Two quantities must be determined: radial velocity and proper motion. The first concerns the motion in the line of sight, creating a Doppler Effect (shift of the spectral lines). The second is the change of the star position, projected on the sphere. Measuring both, you get the true space motion by simple vector addition. By travelling almost in the same direction, the motions of cluster members are correlated, whereas nonmembers perform a random walk. This fact allows distance measurements, known as 'star stream parallax'. Of course, for nearby clusters, the distance of its members can be determined directly by the trigonometric parallax. If the distance is known, the absolute brightness (luminosity) of the cluster stars can be calculated.

Open clusters are an ideal test laboratory for the theory of stellar evolution. The reason is simple: all stars have almost the same age, composition and distance. The main difference is the mass. Thus, the physical state of the cluster is controlled by a single parameter – a perfect situation for astrophysicists. Mass determines both the luminosity and the age of a star. The more massive a star is, the shorter its lifetime. Sunlike stars fuse hydrogen to helium in the hot core. If the fuel runs out (for the Sun this happens in about 4 billion years), the star inflates to a red giant, 100-times larger than the Sun. The core shrinks and the temperature rises. At about 200 million K, helium is spontanuously fused into carbon, nitrogen and oxygen (helium flash). By this event, the outer layers of the star are blown into space. The result is called a planetary nebula with a white dwarf in the centre.

If the initial star was more massive than the Sun, the end is much more dramatic. First, the star is now able to produce heavier elements beyond carbon, nitrogen and oxygen. Unfortunately, the sequence stops with the appearance of iron, when the core temperature reaches 10 billion K. Meanwhile the star has inflated to a red supergiant, 2,500-times larger than the Sun. Fusion ends with iron, because any heavier element would need energy for its production. Thus, gravitation finally wins the race against radiation pressure and the core collapses. The result is a supernova. During this dramatic event, the star shines as bright as 100 billion suns. The process also creates elements beyond iron up to uranium. If the iron core is not heavier than about 3 solar masses, it will survive as a neutron star, surrounded by the rapidly expanding hot cloud of gas (supernova remnant). The matter ejected in the final phases forms the basis for new generations of stars – with a slightly different chemical composition. However, if the core exceeds three solar masses, no stellar object remains: the collapse continues to form a black hole.

All this happens during the lifetime of the open cluster. Thus, at each time the members present a certain mix of the different stages of stellar evolution. The relevant tool to show this is the Hertzsprung-Russell Diagram (HRD). It visualizes the relation between luminosity (absolute brightness) and spectral class (surface temperature) for a large sample of stars.

The famous Hertzsprung-Russell Diagramm (HRD) plots the stars according to their luminosity and surface temperature (spectral class). The Sun is on the 'main sequence' = stars fusing hydrogen into helium. The temperature ranges from 3,000 to 30,000 Kelvin and the luminosity from 10^{-4} to 10^6 times the solar value.

There is a diagonal line, occupied by all stars just fusing hydrogen into helium – the 'main sequence'. Of course, the Sun is placed here. The exact position depends crucially on the mass. In the upper left part of the main sequence we find stars of type O or B. They are large, hot and luminous – and live a short life. In the lower right part, the stars are of type M: small, cool and faint – and long-lived. Beside the main sequence, there are other places, occupied by stars in the later stages of stellar evolution when heavier elements are burned, red giants and supergiants (upper right). Star cadavers, the white dwarfs, are in the lower left region. The HRD of an open cluster clearly shows the current state of evolution, i.e. how many stars are still in the phase of hydrogen fusion and which are in later ones. This determines the age of the cluster.

Classification

Very early on, an attempt was made to classify open clusters on the basis of their optical appearance. Robert Trümpler's three-part classification from 1930 has been largely accepted. He distinguishes concentration, brightness distribution, and richness. Type designations such as IV3p or II2rn originate from his work.

Trümpler classification of open clusters (S = symbol)

Feature	S	Meaning
concentration	I	strong concentration, cluster stands out clearly from the background
	II	low concentration, but still standing out clearly from the background
	III	without noticeable concentration against the centre, but still standing out from the background
	IV	gives only the impression of a random distribution in the star field
brightness	1	all stars almost equally bright
	2	even distribution of brightness over a certain range
	3	little bright, many faint stars
richness	p	poor (less than 50 stars)
	m	medium (50–100 stars)
	r	rich (100–500 stars)
	v	very rich (more than 500 stars)
additional	a	asymmetrical shape
	e	elliptical shape
	n	nebulosity involved

The compact open cluster NGC 2367 in Canis Major has the Trümpler class IV3p. From the table, we learn that this is an object that looks like a random cluster of a few bright and many faint stars (less than 50).

Basically, however, the given values can only serve as a rough guide as they are not quantitative and depend much on observational parameters and the cluster environment. Moreover, it should be mentioned that the Trümpler classification was created from photographs which rarely match the visual impression. II3rn refers to a star-rich (100–500), low concentrated cluster with little bright and many faint stars plus nebulosity.

Associations, Moving Groups, and Asterisms

In addition to the proper open clusters, the Milky Way offers other physical ensembles of stars, which prove to be physically linked only after a detailed analysis. The main types are associations and moving groups. They usually provide the observer with a rather unspectacular view.

In 1949, the Russian astronomer Victor Ambartsumian coined the term 'association' to describe a spatial concentration of stars of almost the same spectral class and thus of comparable physical properties. One mainly distiguishes OB and T associations. The former are local accumulations of young, massive O and B stars. In the spatial areas in question, the stellar density of these types is significantly higher than in the surroundings. With diameters of 100 to 600 light-years, OB associations are significantly larger than open clusters. These very young aggregates are often embedded in nebulosity. Due to their low gravitational bond, the lifetime is less than 10 million years. T associations are accumulations of protostars like T Tauri, still being in the contraction phase. There are currently about 100 associations known, located in the spiral arms. The nomenclature follows the scheme: constellation, type of association, serial number; an example is Ori OB1, the first OB association in Orion.

Moving groups show a common (generally high) space motion of its members. On the sphere the stars seem to move to a common fixed point (apex). The celebrated example is the Hyades in Taurus. The apex lies about 3° east of Beteigeuze. The measured star stream parallax leads to a distance of 151 light-years. Interestingly, the brightest Hyades star, Aldebaran, is not a member of the moving group (with 66 light-years it is much nearer). Due to their proximity, moving groups are very difficult to recognize. Often there is an open cluster in the centre. Another example is the Ursa Major moving group. Interestingly, our Sun lies within the group but does not take part in its motion. The members are therefore spread over the entire sky. Over 100 are known, including five bright stars of the Big Dipper.

Star patterns (asterisms) are not physical clusters. The involved stars have different distances and incompatible space motions. Nevertheless, asterisms are growing in popularity; many offer a more attractive view than some open clusters. Eye-catching examples have been already described in seventeenth century by Hodierna and Hevelius. Much later, it turned out that such patterns are chance alignments. The term 'asterism' is not clearly defined and describes arrangements of only a few arc minutes diameter to those that extend over a large sky area, e.g. the Summer Triangle of Vega, Deneb and Altair, or the Southern Cross (Crux). In order to avoid conflicts with constellations, only small patterns, fitting into the field of view of binoculars or a small telescope should be tolerated.

Globular Clusters

Many globular clusters are already visible with a small telescope. So, it is not surprising that, in addition to bright galaxies and open clusters, they were discovered early on, though without realizing their true nature. The first object, M 22 in Sagittarius, was noticed by the German astronomer Abraham Ihle in 1665, more by chance during an observation of Saturn. The discoverer of M 13, Edmond Halley, noted in 1714 that it 'shows itself to the naked eye when the sky is clear and the Moon does not shine'. However, no single stars could be detected at that time and Messier wrote 50 years later: 'Nebula without stars discovered in the belt of Hercules, it appears round and bright, in the middle brighter than at the edge.'

M 22 in Sagittarius, the first discovered globular cluster.

The Berlin astronomer Johann Elert Bode describes the brightest globular cluster of the northern sky: 'A round and very lively nebula between η and ζ in Hercules. In his position and near distance of two small stars, according to my observation, Mr. Messier sets his diameter to three minutes. In the middle it is brighter than at the edges.' The resolution into single stars and thus the perception as a star cluster succeeded only Hevelius, using his incredibly long 'air refractor'.

For a longer period, the number of globular clusters remained constant. Since the year 2000 a few objects were added, due to observations in the infrared. We currently know of 157. Obviously, compared to open clusters, globular clusters are a pretty rare species.

Structure and Age

The mean stellar density in a globular cluster is about 10 times higher than in an open cluster. For instance, the average distance between stars in M 13 is about 2 light-years. The concentration increases sharply towards the centre by a factor 10. Living there on a planet, one would have anything but a dark night sky. Even the Hubble Space Telescope (HST) could not completely resolve M 15 in Pegasus. An area, only 22 light-years from the centre of the cluster, shows about 30,000 stars.

The masses of globular clusters are quite different, ranging from thousand to a few million solar masses. The extreme case is ω Centauri, hosting about 10 million stars. M 3 in Canes Venatici has 500,000 members. Though M 13 looks more spectacular than M 3, it contains 'only' 100,000 stars. The absolute minimum is AM-4 in Hydra with a population of just over 1,000 stars. As expected, the largest object is ω Centauri with 500 light-years diameter (the smallest are around 65 light-years). The brightest example is 47 Tucanae. Both these southern globulars were initially catalogued as stars.

Globular clusters are very old; they were formed about 11.5 billion years ago. Thus, they are important for the study of the early history of the Milky Way. Globular clusters mainly contain stars of population II: old, metal-poor red giants. This is quite similar to the galactic bulge. At the time when this stellar population formed, the necessary fusion processes for the production of heavy elements had not been set in motion, so the stars mainly consist of primordial matter (hydrogen, helium). However, globular clusters do not only house old stars: There are also young-looking stars known as 'blue stragglers'. For instance, in 47 Tucanae, 20 examples were discovered. These stars, found in the central region, are very hot and consume their atomic fuel much faster than red, cool stars. Since there is no free hydrogen for star formation, blue stragglers can only be explained by the collision and merger of old stars in the dense environment. Recently the situation has become more complicated as a number of globular clusters have been discovered with multiple generations of stars.

However, there are also a few 'young' globulars: Palomar 12 in Capricornus and Ruprecht 106 in Centaurus are about eight billion years old and also much smaller than ordinary exemplars. They might be created later by the interaction of the Milky Way with its companion galaxies like the Magellanic Clouds.

Distribution and Classification

Globular clusters are located in the huge halo around the Milky Way. This knowledge is based on research done by Harlow Shapley of Mount Wilson Observatory. In 1917 he determined the distances of 69 objects using the period-luminosity relation of their RR Lyrae stars (luminous variables with periods of less than a day). Shapley revealed that the globular cluster distribution is not uniform but shows a concentration towards the galactic centre. Thus, many objects lie in or near the constellation Sagittarius.

Globular clusters move on eccentric orbits, which lead them far away from the Milky Way. Among the most distant is NGC 2419 in Lynx, the Intergalactic Wanderer. Its distance (always measured from the Sun) is 269,000 light-years, which is almost twice more than the Magellanic Clouds. The orbital period is three billion years. Such objects in the outback of our galaxy are referred to as 'extreme halo globulars'. The other extreme is M 4 in Scorpius, with a distance of only 7,200 light-years.

The Intergalactic Wanderer NGC 2419 in Lynx is 260,000 light-years away. The star is 4' west and has 7.2 mag.

It was again Shapley who introduced the first classification of globular clusters along with Helen Sawyer. As they mainly differ in their concentration, they defined 12 classes: I (very compact) to XII (very loose). The remote cluster NGC 7006 in Delphinus is of class I. Another highly concentrated object is M 80 in Scorpius (II). The lowest concentration (XII) shows NGC 5466 in Bootes. However, this says little about whether the globular cluster can be visually resolved into single stars. Not the brightness of the fringe stars counts here, but only the brightest stars inside. The Shapley-Sawyer classification does not include transitions between open and globular cluster, like M 71 in Sagitta.

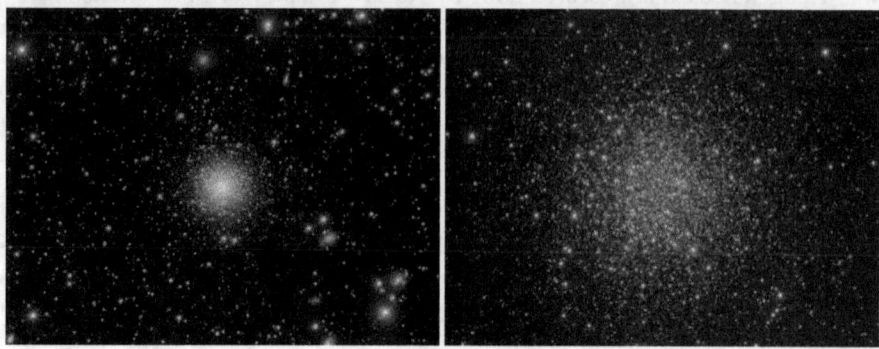

Extreme concentration classes: NGC 7009 in Delphinus (left) is of class I and NGC 5466 in Bootes of class XIII; this low concentrated object is a difficult target.

Extragalactic Globular Clusters

All objects, mentioned so far, are galactic globular clusters. The exploration of extragalactic specimens did not gain momentum until around 1970. So far, such objects have been discovered in over 70 galaxies and it is believed that all large systems (and even dwarfs) have globular clusters as satellites in their halos.

The number of globular clusters is very different in galaxies. Because the Andromeda Nebula has about 300, we suspect a similar value for the Milky Way, but we see only half, the rest are probably hidden behind dark clouds. The unexpectedly large number of 13,000 was found for the giant elliptic galaxy M 87 in Virgo. By contrast, none were detected around NGC 891 in Andromeda. In galaxies with violent star formation, globular clusters were discovered, which do not occur in normal galaxies. They are termed 'super star clusters' or 'blue globulars'. Their special feature, apart from the low age of 10 to 100 million years, is the enormous luminosity and compactness. Super star clusters are huge star-forming regions that probably result from the interaction of galaxies.

Diffuse Nebulae

The first discovery of a nebula – being neither a disguised cluster nor a galaxy – was made in 1610, when the French astronomer Nicolas-Claude Fabri de Peiresc saw the Orion Nebula (M 42) in his small refractor. Hodierna followed with M 8 (1654). The listing of similar objects begins with Messier, who had observed M 1, M 8, M 17, M 20, M 42, M 43 (northen appendix of the Orion Nebula), and M 78. We now know that M 1 is a supernova remnant and M 78 a reflection nebula; the rest are emission nebulae. Later Herschel discovered numerous objects, but he could only sort them by brightness and size. The astrophysical classification began with spectroscopic investigations at the end of the nineteenth century.

The term 'diffuse nebula' refers to the composition (gas, dust) and the hazy, irregular structure. Because these objects belong to the Milky Way, we also speak of 'galactic nebulae' – in contrast to extragalactic nebulae (galaxies). Diffuse nebulae are focused on the galactic plane and often associated with star formation regions.

Diffuse nebulae are divided into self-radiating objects (emitting light), illuminated objects (reflecting light) and dark objects (absorbing light). Thus, we have emission, reflection and dark nebulae. Emission nebulae come in various types, depending on their chemical composition and evolution. Generally, one component must be an excited gas, responsable for the emitted light (here the term 'gaseous nebula' is often used). If the radiating gas is ionized hydrogen, we have a HII region. Another type of emission nebulae, supernova remnants, contain a mix of different elements (many were formed during the supernova event). Strictly speaking, planetary nebulae also belong to the category 'emission nebula', but they traditionally maintain their own one. The form is more regular, often resembling the disc of a planet. Many diffuse nebulae show a mix of emitting, reflecting and dark parts. An interesting case is bipolar nebulae; a special form, called 'cometary', looks like the head and tail of a comet.

Emission Nebulae, Supernova Remnants

This type of nebula emits light of characteristic wavelengths, due to atomic processes. By external influences, the electron in the atomic shell is forced to jump to a higher energy state (excitation) or even to leave it (ionization). About 70% of the interstellar gas consists of hydrogen (H). Ionized hydrogen is called HII, so a nebula containing it is called a HII region. The sources of energy are nearby luminous hot stars of the spectral types O or B, producing ultraviolet light. This leads to a temperature of 10,000 K at the place of the nebula. An ionized hydrogen atom (actually a single proton) soon recombines, i.e. absorbs a nearby free electron. This leads to the emission of light of characteristic wavelengths. Most of the gas is not sphericaly distributed in space, so a HII region usually appears irregular. The boundary of the visible nebula is defined by the place where the radiation of the nearby stars is no longer sufficient for ionization.

In most HII regions, $H\alpha$ is the strongest emission line. It glows deep red and that's why colour images show a reddish nebula. This wavelength is invisible to the eye. Another important line is $H\beta$, appearing blue-green (optimally visible). In addition to hydrogen, emission nebulae can contain heavier elememts like helium, nitrogen, or oxygen. However, these atoms need more energy to emit light (normally by excitation) and therefore can only be found close to very hot stars of spectral type O. A characteristic line is that of double ionized oxygen (OIII).

When there is no star within the critical distance, the hydrogen gas remains neutral (HI). Thus, HI regions are optically invisible. However, they can emit raditation with a wavelength of 21 cm, detected by radio telescopes. Actually, radio astronomy started in the 1950s with the detection of the famous 21 cm line of neutral hydrogen.

Supernova remnants are emission nebulae left behind after the dramatic end of a massive star. The explosion ejects large quantities of the matter into space at speeds of some 10,000 km/s. From the Earth, one observes this spectacle as increasing the brightness of the star by about 20 magnitudes in a few hours. In the last 2000 years, dozen Milky Way supernovae were recorded as 'guest stars'. The extremely hot core of the star can either be destroyed or remain as an ultradense object with a diameter of only 10 km. It consists of practically only neutrons, thus the name 'neutron star', and rotates in milliseconds. This creates a strong magnetic field, which poles are usually not in alignment with the rotation axis. At the magnetic poles matter is ejected as in a gigantic particle accelerator. This creates strong synchrotron radiation (visible light). The existing angle between the rotation axis and the magnetic axis causes a lighthouse effect. The synchrotron beam (jet) circles around and can periodically hit the Earth when the inclination fits – we see a pulsar. Even more massive stars collapse to form black holes.

Reflection and Dark Nebulae

Reflection nebulae are illuminated dust envelopes around bright stars. In contrast to the case of emission nebulae, the visible radiation is not the intrinsic radiation of the nebula, but the scattered light of a nearby star. All these objects therefore show a continuous spectrum. They appear blue, because this wavelength of the starlight is best scattered by the dust; the phenomenon is known as Rayleigh scattering (the same as that which causes the blue sky).

Reflection nebulae are formed around stars that are not hot enough to create an HII region, but bright enough to illuminate the surrounding dust so that it is still visible at a greater distance. This corresponds to stars of the spectral classes B to A. With cool, red stars also reddish nebulae appear, like the faint dust clouds around the red giant Antares. Reflection nebulae can only be observed up to a distance of a few thousand light-years as their surface brightness is considerably lower than that of an emission nebula. The objects are all relatively close to the galactic plane. There are several areas where they appear as visible peaks of a large dark regions, such as around the Pleiades or the Orion Nebula, in Monoceros and Musca/Chaemaeleon.

A remarkable reflection nebula is NGC 1555 in Taurus. The asymmetric object is known as Hind's Variable Nebula. Located near the variable protostar T Tauri, it reacts with changes of brightness and shape. Shortly after the discovery by the English astronomer John Russell Hind in 1852, another variable nebula appeared in the vicinity: NGC 1554, which later was missed (Struve's Lost Nebula).

Hind's Variable Nebula NGC 1555 in Taurus. The densest part is due to reflected light of the variable star T Tau (the brighter star in the middle has 8.4 mag).

Another interesting case can be found in the Pleiades. Photographic images show that the bright open cluster is associated with faint nebulosity. Two objects are catalogued in the NGC: the Merope Nebula (NGC 1435) and the Maia Nebula (NGC 1432). It was long thought that both are reflection nebulae. But in 1995 it was found out that the Pleiades nebulae are not remnants of the matter, which once formed the stars, but dust clouds, currently roaming through the cluster.

Dark nebulae are large, cold interstellar clouds of hydrogen molecules (H_2), water and carbon compounds (dust). These 'molecular clouds' are the raw material for star formation. The objects are only visible through the strong weakening and reddening (extinction) of the light from the stars behind them or when lying against a background of luminous gas. The most important factor for the contrast is the opacity. It indicates, in a scale of 1 to 6, the absorption of the background light. Roughly speaking, the dark nebulae with the largest opacity values are also the easiest to recognize. However, this only applies if the background brightness is sufficient; in front of a bright Milky Way cloud for instance objects with lower opacity are also visible.

Bipolar and Cometary Nebulae

Nebulae that have a clear double structure (like a dumbbell) are called 'bipolar', without considering their emission or reflection nature. This class is therefore not defined primarily physically, but according to the optical appearance. A bipolar nebula has two opposite lobes; often a star is visible at the vertex.

This extraordinary structure refers to an early stage of stellar evolution. Very young stars, mostly luminous objects of the spectral classes O and B, illuminate the dense cloud of dust in which they were born. A strong poleward stellar wind creates bipolar cavities around the star, the walls of which are illuminated from within. Hot stars stimulate the surrounding nebula. The result is a bipolar emission nebula; Sh2-106 in the Cygnus is a typical example. If the energy of the central star is not sufficient, it remains with a bipolar reflection nebula. Of course, these objects differ greatly in their spectral properties. A compact dust torus often surrounds younger objects in the equatorial plane around the star, whose light is then weakened and reddened. Depending on the viewing angle, the two branches appear different in shape and intensity, like for NGC 2163 in Orion.

The bipolar nebula Sh2-106 in Cygnus.

Cometary nebulae are bipolar objects where only one branch is visible or the activity is asymmetric. In combination with the star the object appears like a comet. There are reflection and emission cases. The cometary nebula NGC 2261 in Monoceros is of the reflection type. Not only because of its brightness, it is one of the most interesting deep-sky objects. It was Hubble who detected a changing shape and brightness in 1916. This led to the name Hubble's Variable

Nebula. The nucleus of the 'comet' is the star R Mon, whose variable light, together with dust clouds passing in front of it, causes different illumination effects.

Hubble's Variable nebula NGC 2261, a cometary nebula in Monoceros, illuminated by the variable star R Mon.

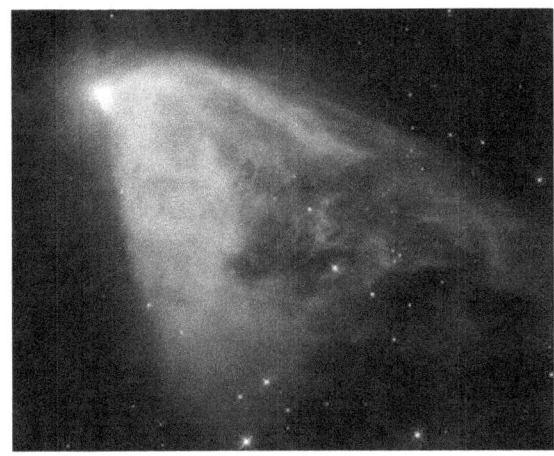

Planetary Nebulae

In 1779 the French astronomer Antoine Darquier saw a nebulous object in Lyra, writing: 'nebula between γ & β of Lyra; it is very dull, but perfectly limited; it is large as Jupiter and resembles a dying planet.' Messier had actually discovered the curious object a few days earlier and catalogued it as M 57. Later observations revealed a ringlike structure and a faint central star. We know it as the 'Ring Nebula'.

Only four planetary nebulae are included in the Messier catalogue: M 27, M 57, M 76, and M 97. Many of these objects appeared in the telescope like dull discs, to be distinguished by Uranus (discovered by Herschel in 1781) only by the lack of motion. This appearance was the basis for Herschel's class IV. He coined the name 'planetary nebula' for these objects in the 1780's. The class contains 78 entries, but only 29 of them are planetary nebulae in the astrophysical sence. Later visual spectroscopy was the method to identify stellar looking objects.

Distribution and Physical Nature

Planetary nebulae are distributed in the Milky Way like the visible stars, with a concentration in the disc. Many objects can be found in the region of the galactic centre in Sagittarius. Only a few are far from the galactic plane, for instance, NGC 4361 in Corvus, with a galactic latitude of 43°.

The planetary nebula NGC 4361 in Corvus has a diameter 2.1' and shines at 10.9 mag. The 13.2 mag central star is easily visible.

As explained the section about diffuse nebulae, planetary nebulae are the product of a dramatic mass loss in the final phase of main-sequence stars with 1 to 8 solar masses. The remaining star is an earth-sized, extremely dense and hot object, a white dwarf. Planetary nebulae are gas clouds, mainly of hydrogen, helium, and oxygen, with diameters of a few light-years. Due to the high-energy (ultraviolet) radiation of the central star, the gas is extremely hot (50,000 to 150,000 K), leading to highly ionized states of the atoms. The nebulous shell expands at speeds of 20 to 30 km/s.

A planetary nebula is a relatively short-lived phenomenon that disappears in about ten to 40 thousand years after its formation. In its early phase one speaks of a protoplanetary nebula, showing processes that are only decades or centuries old. The Egg Nebula in Cygnus is a well-known example. The same applies to very old specimens in which the ejected matter has already been widely distributed in space. On photographic images, they are difficult to detect. In many cases, the asymmetrical appearance of the objects has little in common with a planetary nebula, but rather resembles old supernova remnants. In the final phase it is no longer the original wind from the central star that decides the shape, but increasingly the interaction with the interstellar medium (examples can be found in Abell's catalogue).

Spectrum and Classification

Planetary nebulae show a characteristic emission spectrum similar to HII regions. In the visual domain, the green OIII line of doubly ionized oxygen dominates. In some objects, the colour also changes to bluish, which is due to the influence of the visually second strongest line (Hβ). Old planetary nebulae emit only the red Hα light, barely visible to the eye. As digital cameras are most sensitive in the red, they are priviliged.

Although one might assume that planetary nebulae all look the same due to their genesis, there are many different forms. Although bubbles and rings dominate (historically referred to as 'annular nebulae'), most objects are caused by repeated mass loss from the central stars. These shells, ejected at different times, interact with each other to produce the most complex shapes, such as dumbbells, double discs, bars, or multiple shells. At least 50% of all objects also show an outer, very faint halo. Examples are M 57, M 27 and NGC 6543, which shows a remarkable knot (IC 4677).

The bright planetary NGC 6543 in Draco is called Cat's Eye Nebula. The remarkable emission knot to the west is catalogued as IC 4677. The star at the edge has 9.8 mag.

The Milky Way

Does the fact that we live in a galaxy give us an advantage in revealing the structure of the star system? Yes and no. Of course, you can study the components (stars, gas, dust) at close range, but we are unable to see the big picture. For astronomers, it was very difficult to detect the spiral arms of the Milky Way from within. Accordingly, it took a long time too to recognize spiral nebulae as extragalactic copies of the Milky Way – a quick view from outside would be sufficient, but this is science fiction. William Herschel was the first to come up with an idea about the shape of our galaxy that was not based on pure speculation. His extensive star counts revealed an elongated object (the Sun is not exactly in the centre). However, he chose only a section of the Milky Way. Unfortunately, later his impressive presentation was incorrectly interpreted as showing the whole stellar system.

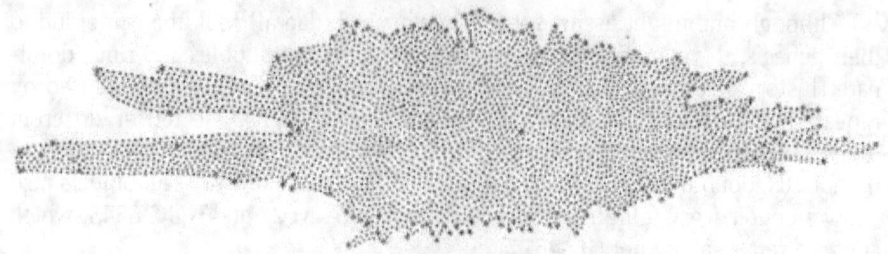

William Herschel's vertical section of the Milky Way (dot right of centre = Sun).

In the 1930s the structure of our galaxy was analysed by statistical methods, using astrometrical data of a large number of stars. This enormous task was done in the visual range of the spectrum. It soon became clear that the results are influenced by interstellar dust, causing a dimming and reddening of the star light. Fortunately, dust is much more transparent to infrared (heat radiation), submillimetre and radio waves. With the corresponding detectors, the structure of the Milky Way was tackled in the following decades. This eventually brought a clear picture of our galaxy – from the inside. Thanks to measurements from the GAIA satellite it is currently much more refined. The European astrometry satellite measures the distances and motions of 1 billion stars down to 20 mag.

However, the outward view is limited too. Hubble had noticed that there is a lack of galaxies in the Milky Way band; he coined the term 'zone of avoidance'. Looking at spiral galaxies in edge-on orientation such as NGC 891 in Andromeda, the reason is clearly seen: a dark lane (due to dust), which apparently divides the disc. In the zone of avoidance, only a few extragalactic objects overcome the severe absorption. With a clear view, objects like IC 10 in Cassiopeia, located at galactic latitude of only −3.3°, would be impressive in the sky.

Compared to other galaxies, the Milky Way is in many ways a normal object. This justifies its meaning as a cosmic standard (the same applies to our home star, the Sun). The visible mass is probably more than 800 billion solar masses. As far as the distribution of the chemical elements is concerned, it consists of 73% hydrogen, 25% helium and 2% metals (all atoms heavier than helium). Looking at the building blocks, one finds 80% low-mass stars – including the Sun – and 10% massive, luminous stars (together about 300 billion); the rest mass is distributed to about 10% gas and 0.1% dust. Though the amount of dust is small, this kind of matter has a significant effect.

The age of the Milky Way is estimated at 12.5 billion years. Actually, it is not an aging diva, but a dynamic object, in which new stars are constantly being born. There is still plenty of interstellar matter like huge molecular clouds, already enriched with metals from previous generations of stars. Currently, about one solar mass per year is converted into stars (in the past, the rate was higher).

Being a prototype for spiral galaxies, the Milky Way offers the characteristic structural elements:

- disc with spiral arms
- bar-shaped central body (bulge)
- centre with a supermassive black hole
- halo with globular clusters

The galactic disc with its spiral arms has a diameter of about 100,000 light-years and a thickness of about 3000 light-years (slightly increasing towards the bulge). The structure shows up in the sky as the classic Milky Way band. Its irregular appearance with bright areas and dark clouds reflects the inhomogeneous matter distribution. Particularly striking are the bright star clouds in Sagittarius and Scutum, the Great Rift in Cygnus or the Coalsack in Crux.

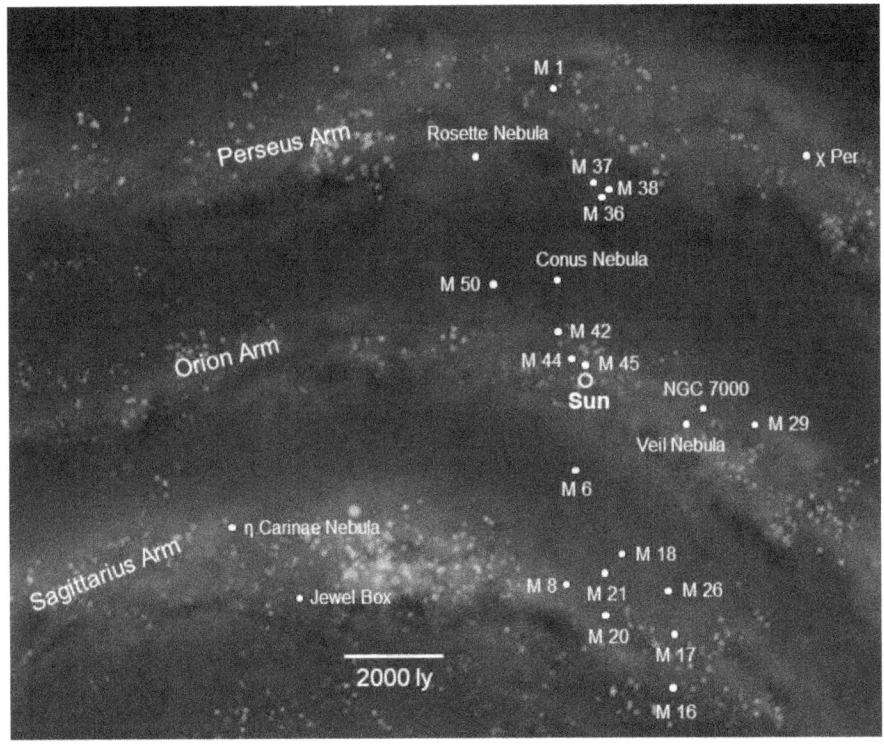

Nearby spiral arms of the Milky Way and its deep-sky objects.

The spiral arms mainly contain young and massive stars (population I). There are two major ones, the Perseus and the Scutum-Centaurus Arm. In addition, there are at least three smaller arms: Orion Arm, Sagittarius Arm, and Outer Arm. The Sun is located in the Orion Arm (also called the Local Arm), about 27,000 light-years from the centre of the Milky Way. In the opposite direction, the edge of the disc is about 13,000 light-years away. Our home region, the Local Arm, and the neighboring arms host several fine deep-sky objects. The following table lists some showpieces, located around us.

Popular deep-sky objects, placed in the Local arm, Perseus arm (further outside), and Sagittarius arm (further inside); distance in light-years.

Spiral arm	Object	Con	Type	Distance
Local Arm (Orion Arm)	Hyades	Tau	OC	151
	Pleiades	Tau	OC	444
	Praesepe (M 44)	Cnc	OC	577
	Orion Nebula (M 42)	Or	EN	1,350
	Dumbbell Nebula (M 27)	Vul	PN	1,400
	Veil Nebula (NGC 6992/95)	Cyg	SNR	1,470
	North America Nebula (NGC 7000)	Cyg	EN	1,600
	Ring Nebula (M 57)	Lyr	PN	2,300
	Owl Nebula (M 97)	UMa	PN	2,500
Perseus Arm	M 36	Aur	OC	4,100
	M 38	Aur	OC	4,200
	M 37	Aur	OC	4,400
	Crab Nebula (M 1)	Tau	SNR	6,300
	Double Cluster (χ Per)	Per	OC	7,500
Sagittarius Arm	Lagoon Nebula (M 8)	Sgr	EN	5,200
	Trifid nebula (M 20)	Sgr	EN	5,200
	Omega Nebula (M 17)	Sgr	EN	5,500
	M 16	Ser	OC	7,000
	η Carinae Nebula	Car	EN	8,000

Disc and spiral structure suggest a rotation around the centre. Physically, a gravitational system of free masses (without external forces) must rotate in order not to collapse. The centrifugal force causes a flattening. Unfortunately, the collective rotation and individual space motions of the stars mix. Fortunately, we could separate the two components with methods working in the near and far. This complex investigation started in the middle of the twentieth century.

In our vicinity, the relative motions of the stars were determined. This needs the measurement of proper motion and radial velocity. The former appears as a slow shift on the sphere; the maximum value is held by Barnard's Star in Ophiuchus with 10.3" per year. For humans, the starry sky remains unchanged – hence the term 'fixed star'. The radial velocity results from the Doppler Effect (shift of the spectral lines). The resulting space motion of stars can have velocities of 100 km/s. The Sun itself moves at 30 km/s in the direction of Hercules (solar apex). The collective rotational motion is tackled by observing interstellar clouds of neutral hydrogen, spread over the whole disc. They emit in the well-known 21 cm line, which penetrates the dust. The motion causes a Doppler shift, measurable with radio telescopes.

Our kinematical investigation in the near and far regions of the Milky Way yields the following result: starting from the centre, the rotational velocity increases linearly (rigid rotation), then changing to a differential rotation. The velocity fluctuates slightly to remain almost constant in the outer disc. At the location of the Sun, the rotational velocity is 220 km/s, resulting in an orbital period of 200 million years (1 galactic year). The rotation curve causes enormous problems, which will be treated a later section.

The disc encloses the bulge, a spherical region with a thickness of 16,000 light-years. It hosts primarily old and metal-poor stars (population II) and therefore resembles in its composition the globular clusters. The bulge stars contain the primordial composition of matter, hydrogen and helium. They are still shining because of their low mass – and the right strategy: use your little fuel sparingly! For a long time, the Milky Way was thought to be a normal spiral galaxy, comparable to the neighboring Andromeda Nebula, which has a spherical bulge. Then in 2005, a 27,000 light-year long bar was detected. Therefore, our home galaxy is now classified as a barred spiral. The main spiral arms start at both ends of the bar.

Illustration of the Milky Way as a barred spiral galaxy.

On the sphere the galactic centre lies in Sagittarius (Sgr). Theoretically, the bar-shaped bulge would extend over 30°. However, most of the light from its stars is absorbed by enormous dust masses. Thus, only a bright concentration of star and dust clouds remains. Nevertheless, under dark sky conditions, it can cast a shadow when standing high in the sky. Astronomical tourists in Namibia report this phenomenon, whereas inexperienced people complain about a 'cloudy sky'.

What we know about the galactic centre comes exclusively from infrared and radio observations. The core region is the strongest radio source of the sky, designated as Sgr A, first detected by the American physicist Karl Jansky in 1932. Later, high-resolution observations, made by combining several radio telescopes (interferometry), brought a spectacular result: the core hosts an extremely compact object, designated Sgr A*, having a diameter of only 22 million km (0.14 astronomical units). This is a supermassive black hole with 4.3 million solar masses. It is orbited by hundreds of giant stars, visible in the infrared. Matter that flows towards the rotating monster initially collects in a hot accretion disc. The final incidence creates high-energy electromagnetic radiation. This and the gravitational influence on the surrounding stars are the main evidence for the black hole. It is probable that every galaxy has such a gravitational monster in its centre. So, again, our home galaxy is nothing special. Compared to M 31 it's even second-rate: our cosmic brother bears a supermassive black hole of about 150 million solar masses.

Looking outside, the Milky Way is surrounded by a spherical halo. With 600,000 light-years diameter, it is six times larger than the galactic disc. The halo is populated by red dwarf stars, which probably emerged already 600 million years earlier as the main structures. It is also home to the globular clusters that move around the Milky Way on elliptical orbits. Though about 150 are known, there might be 200, hidden behind dust clouds. Here the Andromeda Nebula beats us again with about 300 exemplars.

Galaxies

Galaxies are the largest single systems of stars in the universe. They are characterized by a bundle of (interrelated) parameters. Some refer to apparent or purely geometrical quantities like position, distance, brightness, size, or orientation. Others describe the physical proproties of the galaxy, e.g. luminosity (absolute brightness), mass, composition, and rotational velocity. Their determination is difficult – it requires the knowledge of distance. Somewhat intermediate is the classification of galaxies. First introduced as a scheme to describe the morphological appearance, it later could be related with evolutionary processes. All information come from the radiation (light, infrared, ultraviolet, X-rays, radio waves etc.), received by special detectors (ground-based or satellite).

Brightness, Size, and Orientation

These parameters relate to the appearance of the galaxy, as viewed from Earth. Most important is the apparent brightness, normally given for a specific colour (yellow = visual, blue, red). Generally, a distinction is made between integrated and surface brightness. Usually, the apparent brightness is quantified by the integrated magnitude. The surface brightness of a galaxy depends on the apparent magnitude and size, i.e. the large and small diameter (a, b, measured in arc minutes). There are high surface brightness objects like the compact galaxy M 77 in Cetus, and the opposite like Barnard's Galaxy NGC 6822 in Sagittarius, a faint, large member of the Local Group.

For elongated galaxies (a > b), the orientation on the sphere plays a role. It is quantified by the position angle (PA). Of course, elongation can have morphological reasons, like in the rare case of a streched ellipsoidal system. But in most cases, it is a projection effect: a disc-shaped galaxy in an oblique view. The relevant parameter, the inclination (i), is defined as the angle between the symmetry axis (perpendicular to the disc) and the observer. It varies between 0° (face-on view) and 90° (edge-on view). Prominent edge-on galaxies are NGC 891, NGC 4565, and NGGC 5907. From the side, the spiral structure is barely visible. This is much different from above. Examples of face-on spirals are M 51, M 74, and M 101. If the spiral pattern, including internal structures like bright HII regions, is perfectly shown, one speaks of a 'grand design' galaxy. The prototype is M 83 in Hydra.

Different views on spiral galaxies: due to its inclination angle, NGC 891 in Andromeda appears edge-on (left) and M 74 in Pisces face-on. From the side, an absorption band is a prominent feature, while in the front view the spiral pattern appears best.

Classification

Edwin Hubble expanded our understanding of the galaxies as individual objects by introducing a revolutionary classification scheme. It still forms the basis for arranging the various forms and structures. Hubble's famous 'tuning fork' diagram of galaxy types is depicted in his classic book *The Realm of the Nebulae* of 1936. The scheme shows three basic types: elliptical (E), normal spiral (S), and barred spiral (SB). Objects, which do not fit are called irregular (I). The empty vertex of the fork was later filled by a transition type (S0, SB0).

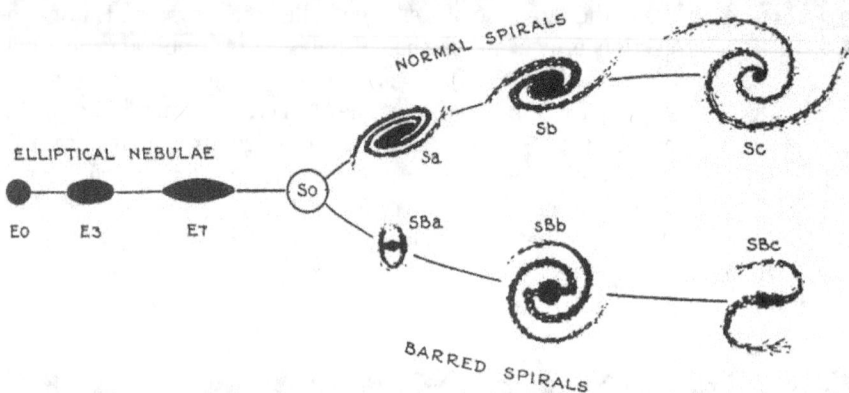

Edwin Hubble's celebrated 'tuning fork' classification system (from his classical book *The Realm of the Nebulae*).

To illustrate the morphological scheme (Hubble type), the book contains images, made with the 60- and 100-inch reflectors on Mount Wilson. Because of the low quality of the rendering, Hubble later decided to make a large-format picture book. Starting from 1948 he photographed typical galaxies with the new 200-inch Hale reflector on Mount Palomar. When the famous astronomer died unexpectedly in 1953, the project was already well advanced. Fortunately, Hubble's successor, Allan Sandage, completed it and published the monumental *Hubble Atlas of Galaxies* in 1961. He also revised the classification scheme and introduced some new designations. Later, Gérard de Vaucouleurs continued this important work.

The elliptical galaxy (E) marks the simplest case. The form is described by the ellipticity, ranging from E0 (round) to E6 (strongly elongated). A nice example of an E0 galaxy is NGC 404 near β And. NGC 3585 in Hydra is of the rare case E6. The ellipticity, like the elongation, is first of all a geometric quantity. It says little about the true shape of the object, which can only be determined by a physical study. One differentiates: prolate (slightly spindle-shaped), oblate (flattened like a thick lens), triaxial (three different major axes), and boxy. Examples are: NGC 741 (prolate), NGC 315 (oblate), NGC 4365 (triaxial), NGC 1600 (boxy).

The elliptical galaxy NGC 3585 in Hydra is of the rare type E6.

Most elliptical galaxies are pretty structureless. The surface brightness steadily increases from edge to centre. A few galaxies show dark structures, due to dusty regions in their body. These peculiar objects usually have a prolate form. Examples are the 'dusty spindles' NGC 5128 in Centaurus and NGC 185 in Andromeda, a faint elliptical companion of M 31. Some extremely elongated galaxies were formerly classified as E7 (no more used). Among them is NGC 3115 in Sextans. The object was thought to be a prolate elliptical galaxy; hence the popular name Spindle Galaxy. Actually, NGC 3115 is of type S0 (lenticular).

NGC 6166 is the central galaxy (type cD) in the rich cluster Abell 2199. It has a multiple nucleus, due to merging smaller galaxies.

63

Elliptical galaxies can be giants or dwarfs. In the former case, we speak of 'cD galaxies', usually found in the centre of galaxy clusters (cD means 'core dominant'). They can have masses of a few billion solar masses (record holder is UGC 12591 in Pegasus); in the core provide supermassive black holes for the necessary energy output. Some objects have multiple nuclei, like NGC 6166 in Hercules. They are cannibals, consuming smaller galaxies in the vicinity. The opposite is marked by dwarf elliptical (dE) or dwarf spheroidal (dwSph) systems. They are probably the infantry of galaxy clusters. Due to their very loose structure they show the lowest surface brightness. The Local Group has many of them.

The dwarf spheroidal Sculptor System is a Local Group member. Due to its extremely loose structure it is a very difficult target.

Normal spirals (S) consist of a spherical central body (bulge) and a surrounding disc with the spiral arms. At least two arms extend from the edge of the bulge. Barred spirals (SB) have a bar-shaped bulge. Two spiral arms start at the ends of the bar. The disc is similar to the S-type. The classification of spiral galaxies uses the following criteria. First, there is the type of the bulge: spherical (S), bar-shaped (SB). The next criterion is the density of the spiral arms, ranging from tight to wide open windings. Five cases are defined: a (narrow), b (medium), c (wide), d (very wide), and m (magellanic). In the latter the spiral pattern is irregular and thus barely visible. The Magellanic Clouds are the defining objects. There are the intermediate cases ab, bc, cd, and dm (the Milky Way is of type SBbc). The following table illustrates the Hubble types S and SB by prominent galaxies.

Examples of spiral galaxies with different Hubble types

Type S	Example	Type SB	Example
Sa	M 104	SBa	NGC 7743
Sb	M 81	SBb	M 95
Sc	M 101	SBc	M 83
Sd	NGC 7793	SBd	NGC 4242
Sm	NGC 4449	SBm	Magellanic Clouds

Edge-on spiral galaxies often show a dust lane, caused by internal extinction; the standard example is NGC 891 in Andromeda, showing an inclination of 88°. Because at this angle there are no spiral arms visible, a classification based on their density is not applicable. But there is an alternative: the bulge-to-disc ratio, calculated from the apparent thicknesses of the bulge and the disc. Sa galaxies have a pronounced bulge, like M 104 (84°). For the type Sb it is smaller, such as in NGC 4565 (86°). Sc galaxies have only a weak bulge, e.g. NGC 5907 (87°). Galaxies of type Sd are pure discs, looking flat in the edge-on view. The most extreme case is the superthin galaxy. Examples are NGC 100 in Pisces and NGC 5529 in Bootes. Finally, how to detect a bar in the edge-on case? Fortunately, the spiral arms sometimes hang above or below the disc, as in the case of the SBc galaxy NGC 7640 in Andromeda. IC 3074 in Virgo is a very flat edge-on galaxy of the rare type SBd.

The very flat edge-on galaxy IC 3074 in Virgo is of the rare type SBd. The 14.1 mag object is a member of the Virgo Cluster.

S0 galaxies are lenticular systems of high surface brightness. Their disc is structureless, i.e. there are no spiral arms. A typical example is the edge-on galaxy NGC 4111 in Canes Venatici. In the Hubble sequence ('tuning fork'), they are at the branch point, connecting E and S/SB galaxies. Accordingly, this type shares some characteristics with its neighbours. The flattening is lower compared to spiral galaxies; bulge and disc form a unit. Even in the edge-on view, the bulge does not stand out. Face-on, S0 galaxies are difficult to distinguish from spherical galaxies (E0). Sometimes a weak bar (SB0) is present, like in the case of NGC 4371 in Virgo. There are objects showing a boxy centre; examples are NGC 128 in Pisces and NGC 7332 in Pegasus.

Irregular galaxies (I) stay off the Hubble sequence. There are lumious examples like NGC 4449 in Canes Venatici, showing violent star formation, as well as dwarfs like IC 1613 in Cetus (a Local Group member). The Magellanic Clouds were first classified as 'I' by Hubble. As we have seen, both are SBm galaxies. Most irregular galaxies are dwarfs. The Local Group contains many of them (some show a weak elliptical or spiral shape). Beyond we encounter an interesting sample: the nine dwarfs, found by the Swedish astronomer Eric Holmberg in the 1930s; they are designated Holmberg I to IX.

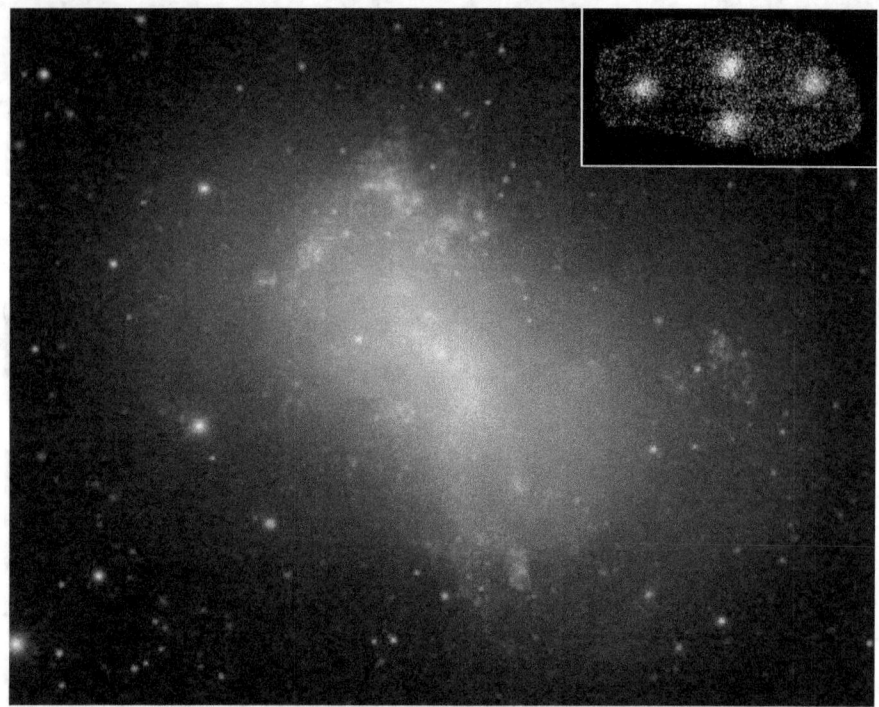

The irregular galaxy NGC 4449 in Canes Venatici is subject to violent star formation. William Herschel has seen only four bright knots (inset).

For the Hubble types it is impossible to get the exact frequency distribution, for we can not classify all galaxies in the observable universe. But we are able to check our cosmic neighborhood, say all NGC/IC galaxies up to a distance of 500 million light-years. In this sample, spirals clearly dominate. Second best are elliptical galaxies. The amount of irregular systems is very small. This was different in earlier times, as we will see.

Statistics of Hubble types on the basis of the NGC/IC galaxies up to a distance of 500 million light-years

Galaxy type	Frequency (%)
elliptical (E)	18.1
lenticular (S0)	14.9
normal spiral (S)	35.5
barred spiral (SB)	29.6
irregular (I)	1.8

Deviations from the norm are relatively common in galaxies. The term 'peculiar' was introduced to draw attention to exceptional structures, seen in galaxies of standard type (E, S0, S, SB). Usually, 'pec' is added to the designation. Prominent examples are M 82 (Sd pec), NGC 128 (S0 pec), NGC 520 (Sa pec), NGC 1275 (S0 pec), or NGC 5128 (S0 pec). The term was introduced by the self-willed American astronomer Halton Arp in his famous *Atlas of Peculiar Galaxies*, published in 1966. The work is based on deep images, taken with the 200-inch on Mount Palomar. The optical peculiarities are often related with a strong radio source. The first detections were designated by the hosting constellation. We have, for instance, NGC 5128 (Centaurus A), NGC 1275 (Perseus A), M 77 (Cetus A), M 87 (Virgo A), or the remote, dusty cD galaxy Cygnus A.

The peculiar lenticular galaxy NGC 128 in Pisces (note the boxy centre). It forms a compact trio of 2' size with NGC 127 (right) and NGC 130.

Most peculiar are ring galaxies. They are the result of a direct, central collision of two galaxies. A small elliptical galaxy has chrashed perpendicular onto a large spiral. After penetration, the remainder of the disc is a ring-like structure. Examples are NGC 985 in Cetus or Hoag's Object in Serpens (a perfect ring with a core). Some objects, known as polar ring galaxies, show a combination of spindle and surrounding ring. Here a S0 galaxy has fused with a smaller object. The standard example is NGC 4650A in Centaurus.

Hoag's Object in Serpens is the rare case of a perfect ring galaxy with spherical core.

There are further strange creatures in the galaxy zoo, fitting in no class. Some look like a comet, boomerang, sandwich, dumbbell, or crab. NGC 4861 in Canes Venatici is the best-known example of a 'cometary galaxy' (this has nothing to do with cometary nebulae). The rare object consists of a compact core (luminous HII region) and a diffuse tail (made of stars). UGC 5938 and UGC 5942 in Draco even form a 'double comet'. In this unique case, two comet-like systems with different position angles are 1.5' apart. Given this diversity, it is not surprising that some objects have been misclassified. An example is Abell 76 in Hydra, first catalogued as a planetary nebula it actually is a ring galaxy, 158 million light-years away.

UGC 5938/42 in Draco is a unique pair of two 'cometary' galaxies. The 16 mag objects are only 1.5' apart.

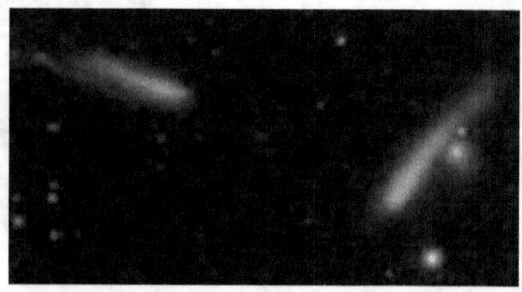

Distance

For galaxies, the distance is often given in Megaparsec (Mpc), where 1 Mpc = 3.26 million light-years. Its determination is a big problem. Assuming that all galaxies have similar extent and luminosity, near objects should appear large and bright, while remote ones small and faint. This would make it easy to get the distance. Unfortunately, diameter and luminosity vary over a large range of values. As for stars, there are giants and dwarfs.

Over the years, the methods of astrometry and astrophysics have succeeded in building up a 'cosmic distance ladder'. It provides a scheme of different distance indicators whose areas of application overlap. Thus, it is essential that a method applied in a certain step must be calibrated with the previous. The first steps of the ladder, where nearby objects are involved, are based on trigonometrical methods (parallax). The following depend on different parameters, like luminosity, diameter, or rotation.

Important distance indicators

Key parameter	Indicator	Remarks
luminosity (absolute magnitude)	Cepheid	period-luminosity relation
	RR Lyrae star	period-luminosity relation
	giant star	
	supernova	detonation of white dwarf (fixed energie)
diameter	globular cluster	
	bright HII region	
internal motion	spiral galaxy	Fisher-Tully relation (rotation velocity)
	elliptical galaxy	Faber-Jackson relation (velocity dispertion)

How we get the luminosity is explained for the classic standard candle: Cepheids (the namesake is the variable star δ Cephei). These are luminous, pulsating stars in the final phase of stellar evolution. They show the celebrated period-luminosity relation: the longer the period, the higher the luminosity. Thus, by measuring the period (usually between 1 and 130 days), the absolute brightness is known. With this method, Hubble realized in the 1920s that nebulae like M 31 are extragalactic systems. Due to the high absolute brightness of Cepheids (up to $M = -5$ mag), the Hubble Space Telescope is able to measure distances to about 100 million light-years.

More far-reaching is the supernova method. Consider a white dwarf, building a close pair with a normal star. By its strong gravity it can suck matter from its companion. If the mass of the white dwarf exceeds a critical value of 1.4 solar masses the object detonates in a supernova (one speaks of type Ia; the collapse of a massive star is type II). Due to the fixed mass, the released energy is equal for all such events, i.e. we always get the same luminosity. The absolute brightness of M = –19.3 mag allows distance measurements up to about 7 billion light-years.

An intermediate indicator is provided by the internal motion of galaxies. To start with spirals, this works in the following way. More mass results in a higher rotational velocity. Since the mass of a galaxy is proportional to its luminosity, the mean rotational velocity correlates with absolute brightness. This is the Fisher-Tully relation. For elliptical galaxies there is a similar result, the Faber-Jackson relation: the brighter the object, the more the scatter of the (random) internal velocities.

Most remote galaxies do not provide a reliable distance indicator (Cepheids or supernovae are rare). Fortunately, we have an ingeneous alternative, based on the redshift. The crucial tool is the Hubble-Lemaître law: $z = (H_0/c)\, r$. It states that the redshift (z), measured in the spectrum of the galaxy, is proportional to its distance (r). H_0 is the proportionality factor (Hubble parameter). It is not a constant since it changes with cosmic time. The index '0' indicates the current value, which is $H_0 = 67$ (km/s)/Mpc.

What causes the redshift? In terms of Einstein's General Relativity, the Hubble-Lemaître law is a consequence of the expansion of the universe. The relative distances between all galaxies increase. Redshift is thus caused by a 'cosmological' Doppler effect, which has nothing to do with a 'recession velocity'. Actually, the expansion of the universe stretches the light to a longer wavelength. Distance measurements, based on redshift, are especially useful in case of large numbers of galaxies. For instance, the multi-object spectrograph at the 4-m VISTA telescope at Paranal Observatory in Chile can measure 4,000 spectra at once. Big redshift campaigns like the Sloan Digital Sky Survey (SDSS) have led to a three-dimensional picture of the cosmic galaxy distribution.

However, there can be peculiar motions of galaxies, causing a classical Doppler shift. Often the gravitational attraction of large masses works against the overall expansion. Thus, the measured redshift is a mix of both effects. A 'local' peculiar motion is the 'Virgo flow', caused by the gravitational pull of the Virgo Cluster on the galaxies of the Local Group. Fortunately, the significance of proper motion in the redshift decreases with larger z. At great distances the smooth 'Hubble flow' (cosmic expansion) wins the race.

Diameter, Luminosity, Mass, and Rotation

Diameter, luminosity, mass and rotation are physical properties of galaxies. Their determination needs the distance to be known. The linear diameter of a galaxy (in light-years) results from its apparent size (a × b) by a simple geometric calculation. As mentioned, luminosity is equivalent to absolute brightness, defined as the hypothetical object brightness at a standard distance of 10 parsecs (32.6 light-years). For the Milky Way we get M = –20.5 mag. For comparison, the Sun has M = +4.7 mag (it would be an insignificant star at 10 pc). Note that absolute brightness and mass share the same letter (M).

The physical parameters of galaxies vary within a reasonable range. The lower end marks spheroidal dwarf galaxies. Segue 2 in Aries, a member of the Local Group, has the lowest values for all parameters. The largest mass has ESO 146-IG5 in Indus, the cD galaxy in the rich cluster Abell 3827. The brightest objects include 'ultra-luminous infrared galaxies' (ULIRG) reaching 10^{11} times the Sun's luminosity. The activity is the result of violent star formation of 100 to 1000 solar masses per year. Quasars can reach M = –33 mag, which is equivalent to 1000 times the brightest galaxies. The largest known galaxy is 5.6 million light-years in diameter: IC 1101 in Virgo, the cD galaxy in Abell 2029.

Range of physical parameters of galaxies in comparison to the values for the Milky Way
(L_S, M_S = solar luminosity/mass)

Quantity	Range	Milky Way
mass	$10^6 \ldots 10^{13} M_S$	$1.8 \times 10^{11} M_S$
absolute brightness	–2.5 ... –33 mag	–20.5 mag
luminosity	$800 \ldots 10^{11} L_S$	$10^{10} L_S$
diameter	500 ... 5.6 million light-years	100,000 light-years

Luminosity (L) and mass (M) of a galaxy are integral quantities, resulting from the sum of the individual objects. Of course, the luminosity increases with the number of stars, thus L is proportional to M. A crucial quantity is the mass-luminosity ratio M/L. If all matter radiates, then M/L = 1 (M, L in solar units). Especially in galaxies with a large amount of non-luminous interstellar matter, the ratio should be slightly larger than 1.

A galaxy is dynamically stable when gravitational force (mass attraction) and centrifugal force (rotation) are equal at each point. The mass distribution, as a function of the radial distance (r) from the centre, can be estimated from the visible matter. This determines the rotational velocity (v). A dynamical calculation then gives a theoretical rotation curve (v as function of r). In the bulge we expect a rigid rotation (v increases linearly). In the inner spiral arms, the mass density is quite constant and so the expected velocity reaches a maximum. To the edge of the disc, the amount of matter gradually drops down to zero. Thus, v should decrease in a non-linear manner, known as Kepler rotation (like in the planetary system).

Fortunately, we can determine the actual rotation curve of the Milky Way, using radio astronomy, measuring the Doppler shifts of the 21 cm line of neutral hydrogen. What is the result? Bulge and inner arms fit well. But in the outer regions the velocity remains constant (at the position of the sun we get 220 km/s), even beyond the edge of the disc, where isolated stars can be checked. Thus, the analysis does not yield M/L ≈ 1, instead, the ratio turned out to be 6!

The difference can only be explained by an enormous amout of additional (invisible) mass. Without it, the outer regions of the Milky Way would fly apart by the centrifugal force at the high speed. It is still unkown what the invisible matter consists of – one only has a name for it – 'dark matter'.

This confusing result was soon confirmed for other spiral galaxies, like the Andromeda Nebula. The required Doppler Effect measurements are easy to do at higher inclinations. They yield that the amount of dark matter must increase towards the edge of the disc (and even beyond), leading to the assumption that it is distributed in a spherical area around the galaxy. One speaks of the corona, which is larger than the (visible) halo; here M/L reaches 100.

Active Galactic Nuclei and Quasars

The term 'active galactic nuclei' (AGN) refers to objects with high luminosity and spectrum with strong emission lines. Examples are the Seyfert galaxies M 77 in Cetus and NGC 4151 in Canes Venatici; the term was introduced by the American astronomer Carl Seyfert. They appear very compact; the core is surrounded by a faint halo. Responsible for these features is a central supermassive black hole.

The larger brother of the AGN is the quasar, which stands for quasi-stellar object (QSO). Much of what is known about AGN also applies (in amplified form) to quasars. The first exemplar was discovered in December 1962 by the Dutch astronomer Maarten Schmidt at Mount Palomar: 3C 273 in Virgo. The 12.8 mag object, indistinguishable from a star, shows a large redshift ($z = 0.158$). Even higher was the value for 3C 48 in Triangulum ($z = 0.367$), with 16.2 mag much fainter. Both quasars were already known by their radio emission; they are listed in the *Third Cambridge Catalogue of Radio Sources* (3C). What is the distance? Unfortunately, the relevant theory, Einstein's General Relativity, give different definitions for cosmological distance. Here we use the 'proper distance'. This is the (not measurable) instantaneous distance at the time of observation. The calculation for 3C 273 and 3C 48 gives 2.10 and 4.65 billion light-years, respectively.

The highest known redshift currently is $z = 11.1$ (galaxy GNz-11 in Ursa Major), giving a proper distance of 32.2 billion light-years. Don't be confused that this is more than 13.8 billion light-years, while the Big Bang happened 13.8 billion years ago and light is the fastest entity. The problem is due to the wrong assumption of a static universe – it actually expands while light of the galaxy is on the way to us. The 'expansion velocity' of space can even exceed the speed of light.

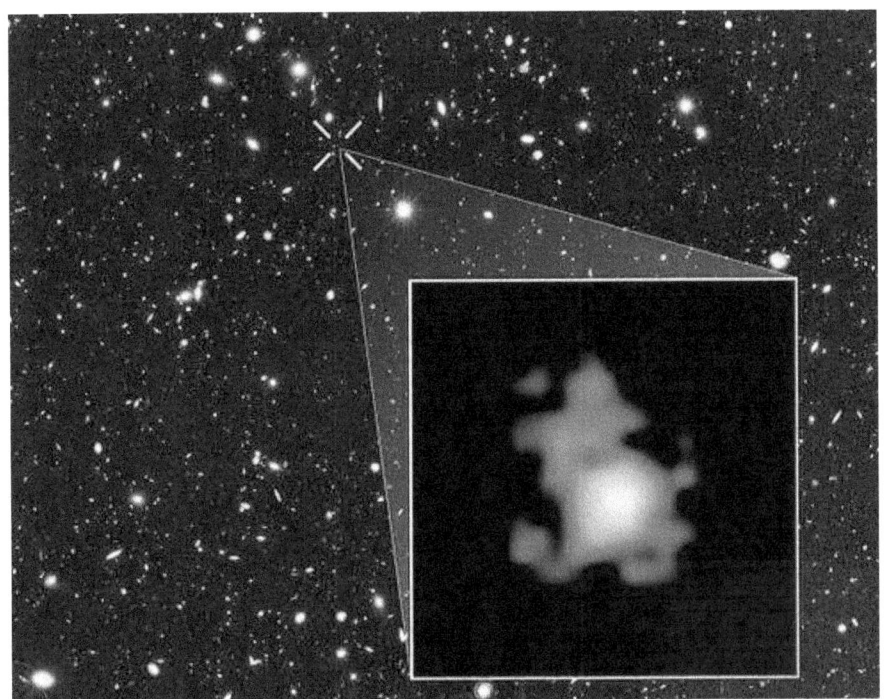

The galaxy GN-z11 in Ursa Major is currently the most distant known object.

Deep imaging has revealed that some quasars (like AGN) are surrounded by a faint halo, belonging to a host galaxy. Thus, a quasar is the extremely luminous nucleus. The supermassive black hole captures enormous amounts of matter. It rotates and is enclosed by an accretion disc that reaches down to the event horizon, which discretely shields the singularity from the rest of the universe. The spinning matter is heated up by friction, causing an intense emission line spectrum. Magnetic fields also play a role, producing jets at the poles. This is (optically visible) synchrotron radiation, recognizable by their polarization. The two opposite jets reach far into space, detectable as huge radio lobes. Particularly well known is the jet of 3C 273, emitted obliquely to the viewing direction.

If one jet is pointed directly in our direction, we speak a BL Lacertae object (Blazar). The namesake is located in Lacerta and was discovered in 1929 as a 'variable star'. A blazar shows only synchrotron radiation, outshining the accretion disc. The lack of emission lines initially prevented the determination of the redshift. Fortunately, the host galaxy of BL Lac was detected in 1974. By carefully shielding the dominant Blazar light, the galaxy redshift could be determined, giving a distance of 945 million light-years.

Most AGN, normal quasars, or BL Lacertae objects are variable. This is due to inconstant feeding of the supermassive black hole. If the central monster catches a whole star, the luminosity can increase dramatically; brightness increases of 5 mag or more are detected. Smaller variations come from consuming sparser food like interstellar matter. When the supply around the supermassive black hole ends, the radiation decreases significantly – the quasar dies. The transition phase is occupied by AGN.

Pairs and Groups of Galaxies

In general, galaxies (similar to stars) tend to form physical pairs or groups. However, there are optical cases too: two galaxies at different distances accidentally lie in the line of sight. Hydra offers a spectacular example: the galaxy 'pair' NGC 3314.

The strange case of the galaxy 'pair' NGC 3314 in Hydra. The objects are at different distances – a spectacular chance alignement.

How can one differentiate between the two cases? Already the appearance gives hints. If the magnitudes are very different, whereas the types look normal, then the fainter partner is probably a background object. There are also two direct proofs for a physical connection. The first is qualitative and bases on a small angular distance between the objects: look for interaction phenomena (deformed spiral structure, tidal tails, matter bridges), caused by close encounters or collisions. In early studies of double and multiple galaxies this was the decisive criterion, the standard example is M 51/NGC 5195 in Canes Venatici. The second proof is quantitative: if the redshifts match, the galaxies must be at the same distance.

The next stages are multiple systems of three to seven galaxies and groups of up to 20 members. There are two structural features: apparent density (number of objects per unit area) and shape (form of arrangement). If the galaxies are very close together, one speaks of a compact group. An extreme example is Shakh 1, discovered by by the Armenian astronomer Romela Shakhbazian in Ursa Major: 20 red stellar objects of 16 mag are confined in a circle of 1.5' diameter. At first, one even suspected a compact cluster of red stars. However, the redshifts testified a compact galaxy group, about 1.3 billion light-years away. The evolution started with a dense swirling group of spirals. By many close encounters the arms were eventually stripped off. What remained were the bulges with their old, reddish shining stars.

Shakh 1 in Ursa Major is an example of a compact group of compact galaxies. The red 16 mag objects cover an area of only 1.5' diameter.

Our Milky Way is a member of a group too, the Local Group. It contains 54 galaxies in a volume of 10 million light-years diameter. Three spiral galaxies dominate: Andromeda Nebula, Milky Way, and M 33 in Triangulum. The rest are irregular elliptical or spheroidal dwarf systems. Most of them are satellites of the Milky Way or M 31. Our largest companions are the two Magellanic Clouds. Significantly fainter are the dwarf galaxies in Fornax and Sculptor. Recent discoveries include the dwarfs in Canis Major, Sagittarius and Aries. With a distance of only 30,000 light-years, the Canis Major system is the nearest galaxy. The Sagittarius galaxy is located at the opposite end of the Milky Way, heavily obscured by the centre. Free in the sky, it would be a magnificent object. The object in Aries (Segue 2) is the tiniest galaxy known. The brightest companions of the Andromeda Nebula are M 32 and M 110 (NGC 205). Beyond the Local Group we find many more galaxy groups, similar in structure and content. An example is the one dominated by M 81 and M 82, about 12 million light-years distant.

Clusters of Galaxies

Galaxy clusters can have several thousand members. The standard catalogue, with 2,712 entries, was compiled by George Abell in 1958. It is based on three criteria: distance, richness (number of galaxies), and compactness (galaxy density). It is interesting that the Virgo Cluster is not contained; it does not meet Abell's third criterion. This catalogue was extended in 1989 by Abell, Corwin and Olowin by adding clusters from the southern surveys to bring the total to 4,072.

A typical rich galaxy cluster has a diameter between 10 and 30 million light-years and contains about 10^{15} solar masses. It remains stable when the gravity balances the centrifugal forces due to the random motion of the members. However, the Swiss astronomer Fritz Zwicky, working on Mount Wilson, noted in the 1930s that the galaxies generally move much too fast. 10 to 100-times more mass is needed to get the observered stability – an issue of dark matter (Zwicky coined the term).

The cosmic hierarchy ends with superclusters. These ensembles of galaxy clusters are the largest known structures in the universe. Actually, we live in one, the Local Supercluster. The centre is occupied by the Virgo Cluster, surrounded by a bunch of smaller clusters and groups of galaxies; one of them is the Local Group with the Milky Way. Superclusters can be identified by statistical methods. Like individual summits point to a mountain range, one also finds arrangements of galaxy clusters, which can not be random.

Superclusters contain up to 20 rich galaxy clusters. The typical diameter is about 350 million light-years. An example is the Coma/A1367 supercluster, whose largest members are the galaxy clusters in Coma Berenices (Abell 1656) and Leo (Abell 1367). The mass is about 10^{16} solar masses, which is 500 times more than the visible one. One of the most massive exemplars is the Great Attractor, 650 million light-years distant. Its gravitational force on the Local Supercluster significantly reduces the 'local' rate of cosmic expansion. Unfortunately, its light must penetrate the Milky Way's zone of avoindance and is strongly weakened. The centre of the Great Attractor is defined by Abell 3627 in Norma, making up 10% of the total mass.

The Evolution of Galaxies and Clusters of Galaxies

The finite speed of light (c = 300,000 km/s) given by nature is a blessing for astrophysics. The light-travel time lets us look into the past. You can see how galaxies are formed and how they evolve in their environment. If $c = \infty$, one would see an instantaneous cosmic situation – a snapshot of the observable universe. The evolution of the objects would then have to be derived from the existing states – a much more difficult task. In the case of stars, one actually faces this situation – the light-travel times in the Milky Way are small in comparison to the age of most stars. The 'size' of the cosmos is therefore an advantage.

Shortly after the Big Bang, happening 13.8 billion years ago, there was only hydrogen and helium. How could galaxies or galaxy clusters form from the primordial matter? In the twentieth century, two contrasting scenarios were developed: James Peebles' hierarchical 'bottom-up model' and Yakov Zeldovich's 'top-down model'. The alternative question: were there first the small objects (galaxies), from which the large structures (galaxy clusters) developed, or was it the other way around? For Peebles, protogalaxies initially emerged everywhere, which then came together to form larger units. For Zeldovich, huge flat clouds of matter ('pancakes') were at the beginning. By random density differences they fragmented by gravity to protogalaxies. This means: aggregation (process runs from bottom to top) against fragmentation (process runs from top to bottom).

Unfortunately, the first computer programs, designed to simulate the formation of cosmic structures, brought frustrating results. Although patterns emerged from the gravitational instabilities in the early universe, the calculated picture bore little resemblance to the observed situation. The structures were not pronounced enough after more than 13 billion years. The reason was quickly found: there was simply too little matter in the model. One had neglected to consider dark matter, which causes the gravitational troughs in which normal matter accumulates over time. However, for this, the dark matter must be 'cold' – a turbulent (thermal) motion would make the desired agglomeration process impossible. This led to the CDM (cold dark matter) model.

The new calculations showed better results but were still unsatisfactory. In 1998, it became obvious that the measured distances of remote supernovae contradict the Hubble-Lemaître law. Everything indicated that there must be a large-scale repulsive force, acting against the expansion. Again, a name was soon found: dark energy. Nothing is known about its physical nature, except that it meets a former concept of Einstein, the obscure 'cosmological constant' (Λ), describing something like a repulsive gravity. This was the missing ingredient for the cosmic simulation. It now is based on normal (baryonic) matter, cold dark matter (CDM) and dark energy (Λ); we thus speak about the ΛCDM model. Starting with plausible initial conditions it makes predictions, which are all confirmed by observations. The new theory – actually a combination of General Relativity and elementary particle physics – is called 'cosmological standard model'.

The standard model explains the

- Big Bang, happening 13.8 billion years ago,
- cosmic abundance of the chemical elements hydrogen, helium, and lithium,
- cosmic background radiation with a current mean temperature of 2.725 K, created 380,000 years after the Big Bang, when the temperature had dropped so much that protons and electrons could combine to form hydrogen atoms,
- structure formation (galaxies, galaxy clusters, voids) from the initial inhomogeneous matter distribution,
- accelerated expansion of the universe.

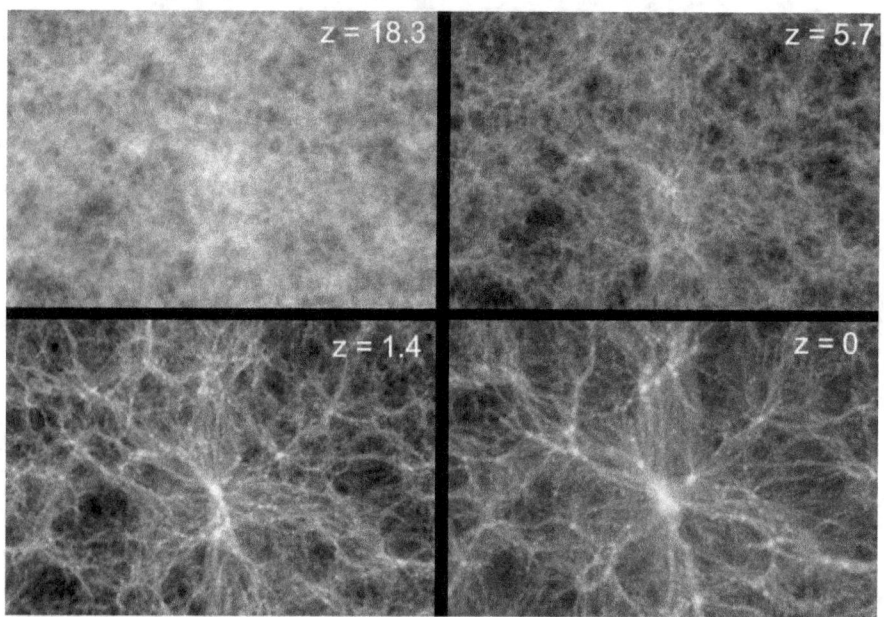

The Millenium Simulation (Virgo Consortium) leads to the observed distribution of matter in the universe. Necessary factors to get the desired result are fixed amounts of normal matter, dark matter, and dark energy.

A milestone was the 'millennium simulation' conducted in the year 2000. It impressively provides the observed image of rich galaxy clusters, long filaments and huge voids (like that in Bootes). It is evident that the inhomogeneous distribution has its origin in primordial density fluctuations, as they are already visible in the cosmic background radiation. They are the seeds of cosmic structures.

Denser regions are the hosts of idle dark matter. Here normal matter was attracted to form protogalaxies, leaving voids at other places. The resulting galaxies formed larger structures, the galaxy clusters, connected by filaments of galaxies and groups. Observation and simulation provide the same picture: a hierarchical structure of the universe, looking much like Swiss cheese – Peebles and Zeldovich were both right somehow.

Looking at a smaller scale, rotating, supermassive black holes of many millions of solar masses have probably played a key role in the formation of galaxies. A stellar black hole, however, needs a star of at least 10 solar masses to form. But even the heaviest specimens, with about 60 solar masses (known from extreme collisions, emitting gravitational waves), are far from sufficient to get a supermassive black hole.

The currently favoured scenario starts with a region having a local concentration of dark matter in which normal matter accumulates, enough to build a new galaxy. For all objects have angular momentum, it comes to a rotating disc-like structure. Because the density is highest in the centre, a million solar-mass portion contracts into a gigantic star, which rapidly collapses into a supermassive black hole. Due to the enormous gravity and rotation it controls the wider environment. The strong forces led to local inhomogeneities and smaller portions of matter gradually condensed into stars – a galaxy is born. The theory is supported by observations of young galaxies whose nuclei are extremely active (quasar phase). There is apparently also a direct connection between the supermassive black hole and the galactic regions with violent star formation.

The famous Hubble Deep Field (HDF) provides a good impression of the early universe. The 2.3' × 2.3' large field in Ursa Major was exposed during 10 days in 1995 with the Hubble Space Telescope. The image shows galaxies, formed just 1 billion years after the Big Bang. There are now many such extreme recordings. All show young disc-shaped galaxies with a lumpy appearance. They evolve into ordinary spirals. The arms are caused by density waves, which spread in the disc. The progressive wave front ensures that the stars approach. They move away when the front leaves the region. The density wave does not really push the stars together, it only changes the choreography of its elliptical orbits around the centre. The ordered twist creates the spiral pattern, which remains stable. Due to the density wave theory the spiral arms are not due to rotation (like a whirlpool). If that were so, they would completely wrap themselves up after a few laps.

Hubble had assumed that elliptical galaxies develop into spiral galaxies. Looking at nearby (old) galaxy clusters, the fraction of elliptical galaxies is 75%. This is significantly higher than in remote (young) ones, where 30% is the rule. In the early universe, there were many more spiral galaxies than there are today. What happened to them? The HDF gives the answer. Many interacting spirals are visible. Apparently, they have reduced each other by merging, forming large elliptical systems, seen in advanced clusters.: Conclusion: spiral galaxies were first and created elliptical galaxies – Hubble was wrong. This is confirmed by computer simulations. The collision of two spiral galaxies is a dramatic event. First, the arms are bent by the strong tidal forces, creating long tails of matter. Gas and dust are compressed, inducing a starburst. Of course, the nuclei with their supermassive black holes are getting closer. They eventually merge to a larger exemplar (the masses simply add).

The Hubble Deep Field South in Tucan was exposed over 10 days in 1998. The part, shown here, has a size of 1.2' × 1.2'. No stars – only galaxies!

Was the Milky Way also affected by mergers? Greater collisions did not happen, but many close encounters. The Magellanic Stream, containing six hydrogen clouds, is a testimony. It is due to the interaction of the Milky Way with the Magellanic Clouds. Evidence also comes from high velocity clouds, falling on the disc, which are likely to be the remnants of dwarf galaxies (the star Arcturus could have come from one). The Sagittarius dwarf galaxy is just being incorporated. Unfortunately, the merger takes place at the opposite side of the Milky Way – otherwise it would be a spectacular view.

However, the big drama will be offered in 3.75 billion years, when Milky Way and Andromeda Nebula collide to form a large elliptical galaxy. This astonishing claim is based on the measured blueshift: currently, the galaxies approach each other with a speed of 300 km/s. When it comes to penetration, the stars are hardly affected by their small size compared to their mutual distances; direct collisions are seldom. They are thrown off their orbits and have to reorient themselves completely. More affected is the interstellar matter. By the collision of large clouds, it is compressed, which strongly stimulates the star formation. Huge HII regions and numerous clusters of young, luminous stars will appear which rapidly end as supernovae.

The Antennae NGC 4038/39, a striking collision of two galaxies.

Will future humans see this spectacle in the Earth's night sky? Certainly not! Due to the Sun's end as a red giant, our planet has long been a lump of ash. Maybe we have found a new home, orbiting another star. Its night sky is covered by a bright spherical star cloud, the image of the giant elliptical galaxy. But the hunger of the new monster is still not satisfied: M 33 and most of the nearby dwarfs will be consumed. Eventually, the (currently boring) Local Group gets its own cD galaxy.

The following table summarises the major steps in the cosmic history, starting with the Big Bang. Looking back, we face an increasing redshift. Note that we currently have detected objects up to $z = 11.1$, being in the epoch of galaxy formation. Of course, the received cosmic background radiation appears with a much higher redshift.

Timescale of cosmic evolution up to now

Event	Years after Big Bang	Redshift
Big Bang	0	∞
Cosmic background radiation	380,000	1,290
End of dark ages (with no stars)	16.5 million	100
First stars	200 million	20
Galaxy halos of dark matter	> 200 million	< 20
Protogalaxies	500 million	10
First supermassive black holes	> 500 million	< 10
First galaxies	> 725 million	< 8
Galaxy clusters	4.3 billion	1.5
Superclusters, voids	6 billion	1
Solar system	9.2 billion	< 0.5

In 3.75 billion years, the Milky Way and our cosmic neighbour, the Andromeda Nebula, will collide. Long before the big clash happens, the Earth's night sky will offer such a spectacular view (NASA illustration).

Practice of Observation

Deep-sky objects can be observed visually or by photography. Though there are technical differences, both methods share some influencing matters, like site, atmospheric conditions and experience. The focus of this chapter is on:

- object selection, observing programs
- telescope, equipment, and optical data
- observation site and atmospheric conditions
- methods of object finding
- technique and subjectivity of visual observation
- textual description, drawing, and evaluation
- technique of astrophotography, camera types

No doubt, an important factor is experience. Newbies should learn from experienced observers. The observational success not only depends on technical subjects, but also on basic physical knowledge about the objects. Otherwise they will be mere spots of light. The enthusiasm comes, for instance, when realizing that the few photons of a quasar, just perceived, are due to an active black hole at the end of the observable universe.

Much more fundamental is the question of clothing, which can not be warm enough, even in summer nights. Coldness, humidity, or wind chill produce stress which kills perception. However, this mainly concerns visual observing in the open air. It is the price an amateur has to pay for the freedom he enjoys. Is life better for professionals?

Until the end of the nineteenth century, an astromomer would still view through the telescope. With the advent of astrophysics, the approach changed dramatically. Especially during the exposure of photographic plates, the night work has focused on mere guiding. The main telescope was occupied by the camera and it was necessary to keep the guide star in the field of a small refractor. Finally, computers completely banished the astronomer from the telescope. Today they have to hand over the expensive instrument with its wide range of additional tools to technicians who are part of the observatory staff; as the danger that equipment will be changed or even damaged is too large. 'Observing' happens on monitors in a control room, shielded from the real sky. But it gets even worse: the trend now is to ban astronomers even from the mountain. The threat has a name: 'service mode'. You wait in your home institute, thousands of kilometers away, until the desired data is recorded and transmitted via the internet. This saves a lot of money. Experiencing the night sky on a lonely peak – once the motivation for many astronomers to take up their profession – is becoming a mere dream.

It follows that as a visually observing amateur you are privileged. This is worth all effort. It might be different for astrophotography, but the professional method leaves its mark: the hobby astronomer sits in a heated room, looking on a monitor to watch the automatic imaging procedure. This sounds like easy observing – but stress can be always near. The computer or the tracking fails, a visitor stumbles over a cable, a smartphone rings...

Object Selection

One can use a clear night to observe ramdomly – just looking around, perhaps with binoculars only. With a smaller aperture, you can scan the sky in a more regular manner, just like Herschel did. Certainly, you will encounter some nice objects. But any serious observing session should be planned. This starts with the selection of objects, visible at the season.

Depending on preference and interest – and, of course, on the performance of the telescope and the available equipment – one can create a list of objects to be looked-up within a certain timeframe. It is important to clarify the following questions in advance. When do the selected objects stand optimally in the sky? At what time does the Moon disturb? The circumstances at the site must also be taken into account (trees or buildings in a certain direction). All these factors yield a relaxed sequence to look-up the targets. Object selection can be based on various sources or ideas:

- sky atlas, planetarium programs, internet databases
- general or object-specific catalogues
- interesting objects
- sky regions or constellations

The classic tool is the printed sky atlas, like the *Uranometria 2000*. It should show deep-sky objects of the standard catalogues (M, NGC/IC), observable with at least 8 inches aperture, and stars down to about 10 or 12 mag. Of course, the new option, whith much more objects, is the digital sky atlas (planetarium program). It brings the real sky to the home PC, laptop, or even the smartphone. High-performance software tools are *Guide*, *TheSky*, or *Eye & Telescope*. They show stars to 16 mag and all deep-sky objects, visible in a 20-inch.

Another option are internet databases like NED or SIMBAD, offering millions of objects. Furthermore, the targets can be viewed via the *Digitized Sky Survey* (DSS); this is the internet version of the *Palomar Observatory Sky Survey* (POSS). An even deeper survey is in the making, The Sloan Digital Sky Survey (SDSS). But be aware that photo and visual impression have little in common. A POSS image (available in blue or red colour) can therefore serve only as a rough guide.

Of course, such huge sources are beyond the scope of the novice. They usually start with the 110 M-objects for which a 4-inch reflector is sufficient. This is easy viewing – success is guaranteed. But there is also a real challenge: the famous 'Messier marathon', viewing all 110 objects in a single night. The only chance for this ambitious mission is in spring. It usually starts at dawn with the galaxy M 74 in Pisces and ends at dusk with the globular cluster M 30 in Capricornus. Note that the marathon is only possible at the right geographical latitude. The northern boundary is marked by the galaxies M 81 and M 82 in Ursa Major at +69° declination. The southernmost object is Ptolemy's Cluster M 7 in Scorpius at −35° declination (M 6 is just 2.5° more north). A small, flexible telescope is the right choice for the task. Of course, experience is urgently needed.

Is the NGC/IC the next step after Messier? Facing the 13,226 targets, one probably gets cold feet. Fortunately, there is an intermediate step: the Herschel catalogues. But again, the whole 2,500 objects might be too much. In this case, the 'Herschel 400' list, created in 1980 by the American Astronomical League, is the right choice. It presents the 400 best Herschel objects, observable in smaller or medium-sized telescopes. When finished, the 'Herschel 2500' is waiting. Here an aperture of 16 inches is required.

Telescope and Equipment

This section concerns essential devices, needed both for visual observing and photography: telescope, mounting, eyepiece, focuser, finder, and filter. For most deep-sky objects, listed in the standard catalogues, a great light gathering power is needed, i.e. a sufficient aperture. Here the Newtonian reflector is standard. Very popular is the low-cost variant, the Dobsonian, created for easy viewing by the San Francisco amateur John Dobson about 1970. With its simple altazimuthal mounting, a large aperture (mirror diameter) can be realized.

Note that, for historical reasons, we use both metric and non-metric units of length in this book: meter (m)/centimetre (cm) and inch ("); it is 1" = 2.54 cm. For reflectors, we find aluminized glass-mirrors of 4 to 20 inches diameter (D) or even more. Their focal length (F) can reach 5 m. We then get a long tube; the eyepiece can only be reached by a ladder – a dangerous task in the night. An important quantity is the aperture ratio (D/F); for a Dobsonian 1:5 is typical. A 4-inch refractor can have 1:10, allowing a high magnification at the expense of a small field of view. However, there are modern glass types (ED) with ratios of 1:4. These rich-field telescopes (formerly called 'comet-seeker') are ideal for large, faint nebulae. Generally, refractors are more expensive then comparable aperture reflectors. This is due to the elaborate optics as well as the common equatorial mounting. A good compromise is a Schmidt-Cassegrain telescope (SCT), with its large aperture (8 to 14 inches are standard) and a focal ratio of

1:10. The long light path is folded here. The compact looking SCT is often used for astrophotography.

An important point is tracking. For astrophotography it is essential and visually it allows relaxed observing, needed to study faint details over a longer time (e.g. for drawing). Of course, tracking is the domain of the equatorial mounting. Only the RA-axis must be moved, either by hand or motor. To follow an object in an altazimuthally mounted telescope, both axes are involved. This is still easy for the Dobsonian but needs experience. Some larger SCTs have sophisticated altazimuth mountings with electronic tracking. Here the field rotation must be compensated. For a Dobsonian such equipment is available.

The focal length of eyepieces ranges between 2.5 and 50 mm. For large values, eyepieces with 2" diameter are a must (instead of the 1.25" standard); a suitable focuser is needed. The achieved magnification (M) depends both on the focal lengths of telescope (F) and eyepiece (f): $M = F/f$. For an 8-inch SCT with 2 m focal length you get magnifications of 40 to 800. For a 20-inch Dobsonian with $F = 2.5$ m we get $M = 50$ to 1,000. Other factors are contrast (ability to show brightness differences) and field of view, measured in arc minutes. For the latter, the viewing angle of the eyepiece is important. A low value (e.g. 40°) produces a 'tunnel view', but with 90° you enjoy a 'space view'. However, now the eye is no longer able to capture the whole scene at a glance – it must 'roam'. Note: good eyepieces are essential – do not save money in the wrong place.

Another important quantity, relevant for visual observing, is the exit pupil. A bundle of parallel rays emerges from the focused eyepiece (actually, a reduced image of the telescope opening). It appears (during the day) as a bright round spot, which does not fill the eyepiece. Its diameter is the exit pupil (p), calculated by $p = D/M$. For a SCT with $D = 20$ cm and $F = 2$ m, equipped with a 40 mm eyepiece, we get $p = 4$ mm. For a Dobsonian with $D = 50$ cm aperture and $F = 2.5$ m, equipped with a 5 mm eyepiece, we get $p = 1$ mm. The relevance of the exit pupil for visual observing is explained later.

The exit pupil appears as a bright round spot, which does not fill the eyepiece.

Filters enhance the contrast in the field of view. There are various types, distinguished by the wavelengths of the light they let pass. The wavelength is measured in nanometers: 1 nm = 10^{-9} m. A broadband filter darkens a lightened sky and objects emerge from the background. A narrow band filter (UHC) brings a significant contrast enhancement for nebulae, emitting in the lines of OIII and Hβ. It is not suitable for galaxies or star clusters, for they consist of stars, emitting light of all wavelengths (continuum). Extreme, and correspondingly expensive, are filters, designed for a single emission line. OIII and Hβ filters are ideal for planetary nebulae or HII regions. The Hα filter is well suited for large (old) planetary nebulae. It is used in astrophotography, because the eye is insensitive at the long wavelength, unless the object is very bright. Thus the Hα filter is the standard tool to visually observe the Sun.

Different types of filters used for visual deep-sky observation (wavelength in nm).

Type	Wavelength	Objects	Remarks
broadband filter	various	all objects	Reduces the light pollution. Brings more contrast for all deep-sky objects, especially galaxies.
narrow band filter (ultra high contrast, UHC)	470–520 and > 630	emission, reflection and planetary nebulae	The band covers the OIII- and Hβ-lines. Well suited as a standard filter. Best for nebulae with a mixed structure. For pure reflection nebulae the effect is small.
line filter (OIII)	495.9 and 500.7	planetary nebulae, special emission nebulae, supernova remnants	From 8 inches aperture, the OIII filter is highly recommended.
line filter (Hβ)	486.1	special emission nebulae	Useful for HII regions.
line filter (Hα)	656.3	large (old) planetary nebulae	Used in astrophotography.

Observation Site and Atmospheric Conditions

The visibility and appearance of deep-sky objects is strongly influenced by atmospheric conditions. Decisive are the turbulence of the air (seeing) and the sky darkness (transparency). In order to be able to recognize small faint objects or fine details, a steady air is necessary. On the other hand, one needs a good transparency to see objects with low surface brightness – a matter of contrast. There are suitable scales to quantify the air quality.

Both factors essentially depend on the observation site, characterized by its local climate, light/air pollution, and altitude. Already Newton was aware that mountain summits are favoured. However, they are subject to strong wind and low temperatures, influencing the observation. A dome is the ideal tool to resist.

Light pollution – the scourge of the astronomer

Air turbulence is due to local density or pressure differences: convection (rising warm air) and advection (horizontal flow = wind). 'Seeing' is the visual effect, most pronounced for bright stars: an erratic variation of brightness, position and colour (scintillation). In the telescope we experience a loss of image sharpness. This can best be tested with double stars. Poor seeing causes faint small objects to disappear, details are blurred. The sensitivity of a telescope against seeing increases with aperture, as now larger turbulent areas influence. With a small aperture you can see through a spatially small and correspondingly less varying air mass. The Greek astronomer Eugène-Michel Antoniadi has defined a simple scale that divides the seeing into five classes.

Antoniadi seeing scale

Seeing class	Condition	Image
I	perfect	stable, without a quiver
II	well	slight quivering with moments of calm lasting several seconds
III	moderate	with larger air tremors that blur the image
IV	poor	constant troublesome undulations of the image
V	very bad	hardly an observation possible

Seeing can also refer to the instrument itself and even the dome. The key issue is temperature balancing. In case of a dome you should bring the inside temperature to the outside value before observing. Also, the telescope is subject to local air flows. One speaks of dome/tube seeing. Concerning temperature balancing, a truss tube (often used for larger Dobsonians) is better than a solid one. Omit humid places; a nearby creek or lake shore produce dew. A dewcap or a small fan can help to get the instrument dry.

A common measure of transparency is the indication of the faintest star (fst), visible to the naked eye. The view of the sky particularly varies with the altitude above the horizon. Therefore, fst-values are often determined near the zenith or in constellations standing high in the sky. Popular is the 'polar sequence'; here all stars up to about 7 mag within 2° of the northern celestial pole are plotted. However, you should first familiarize yourself with the matter. Frequent looking back and forth between the map and the sky (even using red light) makes no sense. Over time, suitable star fields become well-known. Note that the fst-determination can only be carried out when the eye has sufficiently dark adapted. Of course, the eyes must be in a sane condition (already glasses are a hindrance).

Bortle transparency scale

Bortle class	Sky condition	Description	fst mag
1	excellent dark	Zodiacal light, gegenschein and zodiacal band are visible; M 33 is a conspicuous object; the bright Milky Way clouds cast obvious shadows.	7.8
2	truly dark	No stray light; M 33 is easy visible; the Milky is highly structured.	7.3
3	rural	Weak light pollution near the horizon; the Milky Way looks complex; M 33 is easily visible by indirect vision.	6.8
4	rural/ suburban	Fairly obvious light domes over population centres; zodiacal light is still visible; The Milky Way is impressive, showing only the most obvious structures; M 33 is barely visible by indirect vision.	6.3
5	suburban	Light pollution is present in any horizontal direction; zodiacal light is very weak; the Milky Way is dim near the horizon and nearly washed out at the zenith.	5.8
6	bright suburban	Up to an altitude of 35° the sky looks gray; the Milky Way is only visible overhead; M 31 is only modestly apparent.	5.3
7	suburban/ urban	The sky is entirely covered by diffuse gray light; the Milky Way is nearly invisible; M 31 is a difficult object.	4.8
8	city	The entire sky appears in white-gray or orange; a newspaper can be read easily; M 31 is extremely difficult.	4.3
9	inner-city	Bright sky, wherever you look; the Pleiades is the only deep-sky object; only the brightest stars or constellations are visible.	≤ 3.8

An alternative transparency assessment is based on the visual appearance of large naked-eye objects. In 2001 the American amateur John Bortle published a nine-step scale ('Bortle class') referring to the visibility of the Milky Way, the zodiacal light and the galaxies M 31 and M 33. There are only a few sites in the world where a top-class sky can be experienced.

Today, a device called 'sky quality meter' (SQM) is pretty popular, measuring transparency by the surface brightness of the sky in the unit mag per square arc second. Note that even a mountain sky is not really dark. To test this, stretch out the hand – it appears much darker against the sky. The eye can therefore perceive much smaller light differences. Conversely, if one lets the hand reflecting the night sky light, one easily recognizes structures on it.

Object finding

The modern tool to find an object is using GoTo – a sky computer installed at the telescope. If both axes are motor driven, the software can move the tube to any celestial position. You simply have to punch in the object designation (or coordinates). This is the best way to show the deep-sky treasures to a larger audience or to locate a target for photography. No doubt, GoTo saves much time. For non-driven telescopes there is a variant called PushTo.

The sporty variant of finding is called 'starhopping'. Though it is more time consuming, it eventually leads to a great the knowledge of the sky. The method means jumping from star to star and field to field on a certain path to finally reach the desired object. Starhopping works by using star patterns (mini-constellations) and connecting lines on any scale. Any path needs preparation in the form of suitable charts: planisphere for the naked eye or a Telrad (finder without lens), small-scale star atlas for the finderscope (small refractor), large-scale atlas for the low power eyepiece, printed finding chart for the observing eyepiece. Also, a planetarium program, perhaps installed on a laptop, can be used at the telescope, making the issue much faster. It is essential that the monitor can be tuned to faint red light.

Starhopping opens the door to the personal sky – this is visual observing at its best. When repeatedly visiting the same target, the brain learns to find the way without any chart. A starhopping tour is not a fixed road. There are alternative paths and many turn-offs leading to interesting objects nearby. Finally, it is always impressive to visitors when presenting the sky by hand.

However, for the beginner, object finding is associated with some obstacles. The transition from naked-eye viewing to the eyepiece leads to orientation problems. Already in binoculars, the number, brightness and the apparent distance between stars appear different. Fortunately, the view is still upright. The situation is changed in the finder and telescope; the view can be inverted and mirror-reversed. When the beginner does not find any deep-sky object, the astronomical career usually ends pretty soon.

The Technique of Visual Observing

Basically, every visual observer sees something different. As proof one needs to show an object only to different persons in the telescope. For example, in a face-on spiral like M 51, the first perceives a diffused core surrounded by spiral arms and an attachted companion, while a second sees only a featureless round nebula, and a third, however, finds nothing at all (and is accordingly disappointed). The key is experience, coupled with the right observing technique. Another important issue is expectation. A beginner is not prepared to encounter such faint light in case of an object, known from beautiful colour pictures. Over time, you know how nebulae and star clusters look in the eyepiece and can appreciate why they are sometimes better or worse to see. Much depends on the right viewing technique. Of course, we must start with the eye.

The retina is equipped with two types of sensors: cones and rods. Cones are found only in the central area around the optical axis. They are high resolution cells which respond to colour but are not very sensitive. The rods are grouped around the central area of the retina with its cones. They are low resolution cells which respond only to grey shades but are highly sensitive. The rods play the active part in the night, detecting faint colourless light ('all cats are grey by night'). The reign of the cones is the day.

The colour is related to the wavelength. Monochromatic light has only one wavelength, visible as a single colour of the continuous spectrum. If different wavelengths are mixed, you get any other colour, including white light. The latter is an equal mix of red, green and blue. The eye is sensitive between 350 nm (blue) and 700 nm (red). The maximum response depends on the cells: 555 nm for cones (yellow), 507 nm for rods (green). Remember that the rods receive colour as grey shades.

As there is not much colour in the night sky, rods are the crucial receptors for visual observation. If one wants to see faint objects, then indirect (averted) vision is a suitable technique. If you look at something directly (along the optical axis of the eye), the light reaches the (less sensitive) cones in the central area of the retina. Thus, for direct vision, the rods remain outside the view. In order to bring them into play, one has to look past the object. Optimal is an angle of 6° to 16° from the optical axis. In this way one can see sources, which are about 4 mag fainter. The price is the lower resolution. Note also the eye's blind spot, the place of the optic nerve, which forwards the light information to the brain. There are no light sensors here. In order to avoid the blind spot in case of indirect vision, one must always look outward from the nose: with the right eye to the right or with the left eye to the left.

Another ability of the eye is its sensitivity to change. Small brightness variations or the active/passive motion of an object can be recognized. The latter is exploited in the technique of field sweeping. If one wants to perceive a faint object, a slight (perhaps periodical) motion of the tube is often sufficient. Of course, this is especially easy on the Dobsonian. The object (passively) changes its position in the field of view and jumps into perception. If this fails, try a higher magnification. Of course, in case of no tracking, the object naturally moves through the field; by this method, Herschel could discover very faint nebulae. Actively moving objects, like asteroids, are ideal targets for the able eye too.

Rods need some time in the dark to reach maximum sensitivity. The adaptation phase can take up to 30 minutes. Meanwhile the eye's pupil (iris) opens to 6–8 mm (decreasing with age). This diameter is called the entrance pupil (P), corresponding to the exit pupil of the eyepiece (p). If the diameter of the light bundle coming from the eyepiece is larger than the opening of the eye, that is p > P, light is lost (vignetting). By contrast, for p > P the light bundle fits well into the entrance pupil, i.e. the information can be completely processed. However, the eye could consume more light. Thus, p = P is the ideal case. This has consequences for the perception of large nebulae.

The largest perceptible surface brightness is obtained as follows. First, get a maximum entrance pupil by sufficient dark adaptation (say P = 7 mm). Then choose an eyepiece whose exit pupil has the same value (p = 7 mm). For a 20 cm (8 inches) Dobsonian with 1:5 you need a 35 mm eyepiece. Now, take the ratio p/P, which is 1. In case of vignetting we get a worse ratio (p/P > 1). You can make the value even larger, which would change anything for perception. You only need a larger exit pupil p. The maximum is reached for the naked eye (i.e. no eyepiece). Then p is the whole sky. For surface brightness is a matter of object size, you get the curious result: no extended nebula appears brighter in the telescope (of any aperture) than with the naked eye!

Finally, let's treat magnification and contrast. The eye is very sensible to contrast differences. Again, you have to differentiate between extended and point sources. A point source ('star') can not be magnified. However, higher power leads to a darker background. When the constrast increases, the object emerges in the field. For faint, extended nebulae it is best to choose a small magnification. A small aperture will do, offering a large field of view. This can be important: if the object occupies too small a part of the field, too little rods are stimulated and it gets lost in the background. If, on the other hand, the magnification is too high, already a part of the object completely fills the field and there is no contrast at all. Conclusion: visual observing needs a lot of knowing – and self-criticism, as we will learn now.

Subjectivity, Description, and Drawing

Visual observation is naturally subjective. This particularly has affected observational astronomers in former centuries. They were confronted with unknown objects whose physical nature could only be speculated. Especially mysterious were nebulae that could not be resolved into stars even with the largest telescopes (many of them are galaxies). The observers were unable to rationally process and classify what they saw. The perception can differ significantly from reality. Foolish images sneak in – especially when looking on an object too long (with high magnification and small field of view). Thus, controversial and mostly fruitless discussions were made about the appearance and nature of nebulous deep-sky objects.

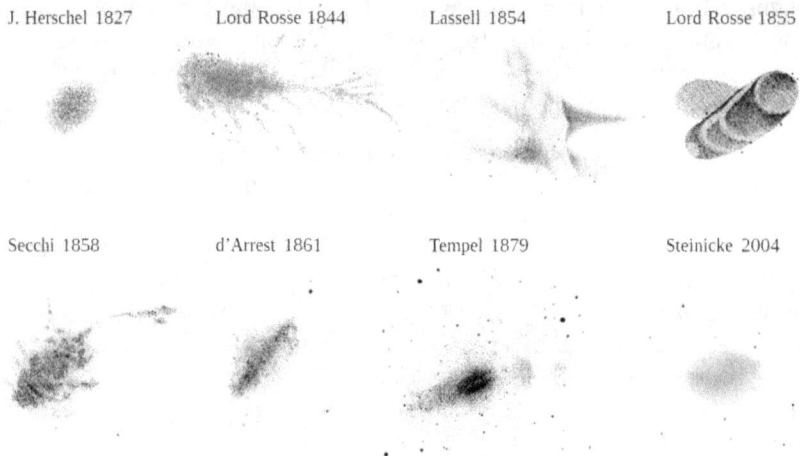

Drawings of M 1 by different observers show a variety of features – which are real?

One would think that these problems are overcome today but that is not the case. It still applies that objects seen for the first time can be misinterpreted. The problem has even worsened, due to 'pretty pictures'. These widely published, mostly coloured images can significantly influence the visual observation. It is a fact that known structures are perceived much more easily than unknown ones. In the worst case you can exactly see what you want to see! Examples are 'obervations' of faint spiral arms, tidal tails in interacting galaxies or very faint quasars – all well beyond the power of the used telescope and the eye. Anyone who 'sees' such impossible entities, experiences the fatal influence of a biased brain. Only one thing can help: a sufficient portion of self-criticism!

Any observation, even a usuccessful one, should be documented – or is forever lost. The following information is worth to note, either at the telescope or later:

- date, start/end of observation
- telescope (type, aperture, focal length)
- observing site, altitude (air/light pollution)
- seeing, transparency, local climate (wind, temperature, dew)
- eyepiece (focal length, magnification, exit pupil, field of view), filter
- observation technique (direct/indirect vision, field sweeping)

To describe an object, one should note: brightness, size, shape, structure, resolution, involved stars, nearby objects etc. A talented observer can make a drawing. This requires sheet of paper, tray, pencil, and light. A headlamp is particularly advantageous. In any case, you should use dim red light; it keeps your dark adaptation (a valuable good). Helpful is a template with a circle that represents the field of view. One should immediately mark the orientation (north, east) by two arrows and enter the diameter (arc minutes). In case of an equatorial mounting, the axes define the directions. Things are not so easy for the altazimuthal Dobsonian. To determine north, one tries to easily pivot in the polar direction. West results from the sky motion (objects drift westward). The field diameter can be determined from a map (comparison of star positions) or calculated from the eyepiece/telescope data. Your drawing should start with the brightest field stars and then sketch, relative to it, the outline of the object. Here you fill in the observed structures. Other objects in the field should be plotted. Drawing from memory (usally on the next day) is not recommended. Warning: do not 'enhance' your drawing by things you have seen elsewhere. Also, digitally processed drawings are worthless. Try to follow the great deep-sky drawers of former times, like John Herschel, Lord Rosse, Wilhelm Tempel, or Étienne Trouvelot.

Astrophotography

The photography of deep-sky objects started in the middle of the nineteenth century. The early (slow) plates were later replaced by more sensitive emulsions. Their shorter exposure time was essential for the monumental *Palomar Observatory Sky Survey* (POSS), based on images taken with the 48-inch Schmidt camera, and the high-resolution images made with the neighboring 200-inch Hale reflector. Another revolution came with the CCD camera. Here, the light is picked up by a planar quantum detector (chip) and converted into electrical impulses, proportional to the intensity. The resulting image is black/white and consists of individual halftone dots (pixels). With the superior sensitivity, faint objects or fine details can be displayed – a blessing for amateurs. Today, the POSS limit (about 21 mag) can already be reached with a 4-inch reflector, equipped with a standard CCD camera. A technical simplification offers the DSLR camera (digital single-lens reflex camera). Thanks to the developments in digital image processing, the amateur is now able to produce colourful shots that look professional in every respect. However, working on the computer is more time-consuming than the actual exposure. Finally, photography needs exact tracking of the telescope. Here the digital camera can help too.

Imaging vs. Visual Observing

Facing the produced results, digital astrophotography seems to be superior to visual observation, though both have to struggle with seeing and transparency at the observing site. However, the two approaches are hardly comparable – each has its own appeal. The major difference: photography is based on accumulating photons (the longer the exposure time, the brighter the image), whereas the eye is not a collector. Our human sensor only shows the current situation ('exposure time' about 1 second). But this does not always have to be a disadvantage.

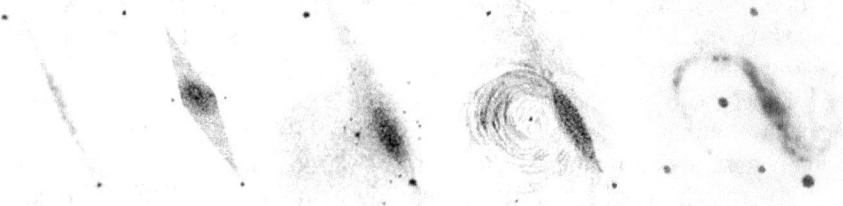

Comparison of various drawings of the barred spiral NGC 7479 in Pegasus, based on visual observation, with the first image, taken by James Keeler in 1899 with 36-inch Crossley reflector at Lick Observatory (right). The four drawings were made by J. Herschel, d'Arrest, Tempel, and Lord Rosse.

It is interesting to see what you can still perceive and where the limits of photography are. To illustrate this, a low surface brightness object and a point source are compared. Visually, a small aperture and low magnification delivers the best results for a faint extended object. For imaging, a large aperture ratio and long exposure time is essential. For perceiving a faint point source, you need a large focal length, allowing high magnification. This is the same for photography, but now a long exposure time is essential. The problem: small details of an object (or a close double star) can not be resolved on the image, whereas the fast eye can detect them in moments of best seeing. Moreover, where visual observation is still possible under sub-optimal skies, digital imaging can not be performed without special tricks (such as the use of narrow-band filters). Without filtering, the sensors are simply too sensitive and do not distinguish between good and bad light (object and background).

Another problem of the digital camera is the inevitable thermal noise in the detector, so that long exposure times are problematic. If the signal-to-noise ratio is sufficiently high, however, summing up several shorter exposures will show even the faintest areas of a nebula, which remain hidden from the eye (using a telescope with the same aperture). Conversely, bright, central details often appear overexposed (saturation). Under certain circumstances, the eye will be able to get a better view here. However, this lack of digital technology can be corrected by overlaying many very short exposures and subsequent image processing. Images of impressive dynamics result. The CCD is most sensitive for red light (in contrast to the eye). For nebulae emitting in Hα, photography has a clear advantage here.

Like the visual observer, the photographer also has to be self-critical because digital image processing has opened up undreamt-of opportunities for conscious or unconscious manipulation. Unfortunately, the international competition for the best 'pretty pictures' encourages this – often at the expense of reality.

Camera Types

The CCD camera is installed in the focal plane of the telescope. It requires a nearby computer, which takes over the device control and often the tracking. To avoid any star trails, tracking is absolutely necessary, even for short exposures. In the case of an equatorial mounting only the RA axis is electronically driven. The altazimuthal mounting is more complex. You need three motors: for both axes and to compensate the field rotation. The classic tracking method uses a second CCD camera as a guider, mounted behind a small refractor at the telescope. In the off-axis method, the telescope light is split with a prism into two perpendicular beams. One reaches the CCD camera, the second the guider. But there are also CCD cameras with a built-in tracking chip, installed next to the main one (on-axis method).

The sensor of the CCD camera must be cooled (the standard temperatures are -10 to -30 °C). This significantly reduces the thermal noise and allows longer exposure times. Another point is the focus. Because there is no screen, the adjustment is pretty difficult. You often have to approach the optimum focus step by step.

The monochromatic chip delivers a greyscale image. To get colour, three images in the standard colours red, green, and blue are taken one after another (not necessary in the same night). The individual intensities were produced using accompanying filters (R, G, B). The colour is created on the computer by putting the three images in different channels; an easy task for the image processing software. The result can be enhanced by a fourth channel, the luminance (L). It provides the best signal-to-noise ratio and ultimately brings power into the picture. As a rule, several short exposures are made for each channel and later summed up digitally. The total exposure times in the channels should match. The result is called LRGB image. Its creation requires a lot of experience and is time consuming. The colour balance is a delicate point; the object should look as 'natural' as possible. Though a certain standard has been developed – particularly used for Hubble Space Telescope images – no one can say how a galaxy looks in colour in 'reality'. The visual impression is pretty worthless.

The DSLR camera is much more comfortable, not only because it has a chip with sensors for the three colours. The device can be used freely at the telescope or the lens is removed and the light is directly focused on the chip. Many models offer a 'live view' on the camera screen. Focusing is as easy as with normal photography. The fact that the red/infrared sensor is almost twice as sensitive as that for green and blue is not a problem for deep-sky imaging. More trouble makes the uncooled chip whose thermal noise limits long exposures. One way out is to reduce the sensitivity (ISO value), which, however, increases the exposure time. It is better to make a number of short exposures. This allows to trick both seeing and tracking errors. We speak of 'lucky imaging': the sharp images are taken, the blurred discarded. Afterwards, the lucky ones are digitally stacked. Even without a telescope, the DSLR camera is ideally suited for direct sky shots with high dynamics. M 31 and M 33 can even be imaged without tracking. For fainter, but not too small deep-sky objects a medium telephoto lens (e.g. F = 500 mm) is ideal. The equipment is mounted 'piggyback' on the tracking telescope.

The luxury class of digital cameras is defined by chips in the classic 35 mm film gauge (36.0 mm × 24.7 mm). At 2 m focal length, the field is 62' × 42'. Concerning resolution, the linear pixel size is important. With 4008 × 2672 pixels, a 35 mm format chip has pixels of 9 μm edge length. At the mentioned focal length that gives 0.9" per pixel (1.8" at 1 m). When recording, you should compare resolution and seeing. With high theoretically achievable resolution, so some recording configuration is worthwhile only with extremely good seeing.

For calibration a dark frame and flatfield should be taken. The dark frame is formed when the shutter is closed (at the same temperature and with the same exposure time as the raw image). The dark current and defective (hot) pixels are contained both in the dark frame and the raw image and can thus be eliminated. For the flatfield, the telescope is directed against a homogeneously illuminated artificial surface. The generated image shows the different sensitivities of the individual pixels, dust and the vignetting of the optical system. In sum, this produces an inhomogeneous image brightness. If you have taken both auxiliary images, the dark frame is first digitally subtracted from the raw image, then the result is divided by the flatfield.

Collection of Interesting Deep-Sky Objects

For practical use, a number of deep-sky objects, located in both hemispheres, are presented here. They are chosen due to their brightness or size, or because the case is simply interesting for various reasons. The scala ranges from spectacular star clusters and nebulae, tiny planetaries, fascinating galaxies to extremely remote galaxy clusters and quasars. A variety of instruments, from the naked eye to a 20-inch, comes into play. The objects are suitable both for visual observation and photography.

Open and Globular Clusters

Open clusters are attractive objects, standing out through their richness, special shape, surrounding nebulae or nice double/multiple stars. No other type offers such a large proportion of objects, already visible with small telescopes. They are advantageous, because a larger field of view and a lower magnitude limit can make the cluster stand out more clearly. With larger apertures as well as on long-exposed images, clusters literally drown in the swarm of background stars.

A magnitude is not reliable for open clusters; it says little about the visibility. Half a dozen 8 to 9 mag stars bring it to a stately overall brightness but in a star-rich area the object is barly detectable. Available information on size and number of stars can also be understood only as a rough guide, since both values are highly dependent on the telescope used or are known only after detailed analysis. That's why even a planetarium program is not very helpful.

The Messier catalogue contains 27 open clusters. Many more can be found in the catalogues of William and John Herschel. The NGC contains 674 open clusters, the IC another 38. Most objects were discovered visually; a few IC clusters were found on plates, looking inconspicuous. Some objects could not be confirmed as physical clusters. Curiously, some large, bright open clusters, like the Pleiades (M 45), Hyades (Mel 25), or the Coma Berenices Cluster (Mel 111) are not in the NGC. They were too obvious for Dreyer to catalogue them. He changed his mind in the IC, including, for instance, M 25 (IC 4725). There are 263 open clusters with NGC/IC designation in the Large Magellanic Cloud (LMC) and 35 in the Small Magellanic Cloud (SMC). The brightest is NGC 2055, belonging to the LMC.

The following table lists some interesting open clusters. Though not a very conspicuous object, NGC 188 is located only 5° from the northern celestial pole. Some clusters are resolved for the naked eye (Pleiades, Hyades, Perseus, Coma Berenices). In other cases, like Praesepe or the Double Cluster in Perseus, the object looks nebulous. IC 2391, NGC 2232, NGC 2264, and NGC 2362 extend around bright stars. Others offer showy star patters, like in M 103.

The compact open cluster NGC 2362 in Canis Major (diameter 6') is dominated by the 4.4 mag star τ CMa – a nice target for small telescopes (use higher magnification).

There are nice chains of open clusters: M 36, M 37, and M 38 in Auriga (length 6°), M 35, NGC 2158, and IC 2157 in Gemini (1°), or the conspicuous case of NGC 6231, Collinder 316 and Trümpler 24 in southern Scorpius (1°). The trio forms a naked-eye sight that resembles a comet. Some people see NGC 6231, discovered by Hodierna in 1654, as the head of the comet; others see the bright pair ζ^1 and ζ^2 Sco, located 30' south of NGC 6231, as head or at least as the leading core of the head. M 11, M 67, NGC 2362, NGC 4755, and Caroline's cluster NGC 7789 are rich and compact objects, offering a fantastic view in a small telescope. The Wild Duck Cluster (M 11) surprises with a red star of 8 mag, standing out near the centre. Some objects are connected with a diffuse nebula: M 16 with the large nebula IC 4703 (Eagle Nebula, containing the famous Pillars of Creation), NGC 2264 with the Cone Nebula or the Pleiades with the nebulae around Maia and Merope (NGC 1432/35). The open cluster M 46, located 1.3° southeast of M 47, even comes with a planetary nebula (NGC 2438), visible in a 4-inch telescope.

The 'false comet' in Scorpius is a chain of open clusters, starting in the south with NGC 6231 and stretching northwards over 1.5°. Below is the bright pair ζ^1 and ζ^2 Sco.

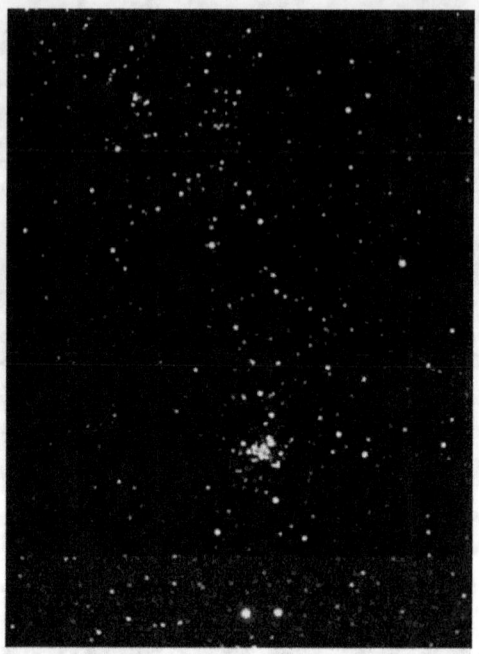

The bright open cluster M 46 in Puppis offers something special: the 10.8 mag planetary nebula NGC 2438. Both objects are not physically related. M 46 is 5,400 light-years away, while NGC 2438 is, with a distance of 2,900 light-years, much nearer.

Interesting open clusters (m = visual magnitude, D = size in arc minutes)

NGC	M	Con	Position	m	Type	D	Remarks
188		Cep	00 47 30.0 +85 15 30	8.1	II2r	15	C 1, most northern open cluster, pretty loose
581	103	Cas	01 33 23.0 +60 39 30	7.4	III2p	6	nice star pattern
752		And	01 57 35.0 +37 50 00	5.7	III1m	75	large, many star chains
869		Per	02 19 04.0 +57 08 06	5.3	I3r	18	Double Cluster, χ Per, naked eye, very rich
884		Per	02 22 05.0 +57 07 48	6.1	I3r	18	Double Cluster, χ Per, naked eye, very rich
1039	34	Per	02 42 05.0 +42 45 42	5.2	II3m	25	some double stars
		Per	03 22 06.0 +48 37 00			184	Mel 20, Perseus Cluster, naked eye
		Tau	03 47 00.0 +24 07 00			109	Pleiades, Mel 22, naked eye, contains Merope and Maia Nebula
		Tau	04 27 00.0 +15 48 00			360	Mel 25, Hyades, naked eye
1647		Tau	04 45 42.2 +19 07 09	6.4	II2m	40	near Hyades
1912	38	Aur	05 28 43.0 +35 51 18	6.4	II2r	15	in chain with M 36 and M 37
1960	36	Aur	05 36 17.7 +34 08 27	6.0	II3m	10	
2099	37	Aur	05 52 18.3 +32 33 11	5.6	I1r	15	very rich
IC 2157		Gem	06 04 50.5 +24 03 58	8.4	III2p	5	in chain with NGC 2158 and M 35
2158		Gem	06 07 25.6 +24 05 46	8.6	II3r	5	
2168	35	Gem	06 09 00.0 +24 21 00	5.1	III2m	25	
2232		Mon	06 28 01.1 −04 50 51	3.9	IV3p	30	around 11 Mon
2264		Mon	06 40 58.2 +09 53 44	4.1	IV3pn	40	Christmas Tree Cluster, around 15 Mon, contains the Cone Nebula
2287	41	CMa	06 46 00.0 −20 45 24	4.5	II3m	39	4° south of Sirius
2323	50	Mon	07 02 47.8 −08 22 33	5.9	II3m	15	
2362		CMa	07 18 41.4 −24 57 15	3.8	I3p	6.0	compact, around τ CMa
2422	47	Pup	07 36 35.0 −14 28 47	4.4	III2m	25	near M 46
2437	46	Pup	07 41 46.8 −14 48 36	6.1	III2m	20	contains planetary nebula NGC 2438
2548	48	Hya	08 13 43.1 −05 45 02	5.8	I2m	30	

IC 2391		Vel	08 40 18.0 −52 55 00	2.6	II3p	60	o Velorum Cluster, naked eye
2632	44	Cnc	08 40 24.0 +19 40 12	3.1	II2m	70	Praesepe, Beehive Cluster, naked eye
2682	67	Cnc	08 51 18.0 +11 49 00	6.9	II2m	25	rich and compact
3293		Car	10 35 51.0 −58 13 48	4.7	I3r	5	Gem Cluster, many brighter stars
IC 2602		Car	10 42 56.5 −64 23 39	1.6		100	Mel 101, Southern Pleiades, naked eye
		Com	12 25 06.0 +26 07 00	1.8		275	Mel 111, Coma Berenices Cluster, naked eye
4755		Cru	12 53 39.0 −60 21 42	4.2	I3r	10	Jewel Box, compact, naked eye
6231		Sco	16 54 10.8 −41 49 27	2.6	I3p	15	naked eye, trio with Cr 316 and Tr 24
6405	6	Sco	17 40 20.0 −32 15 30	4.2	III2p	33	Butterfly Cluster, 2.5° north of M 7
IC 4665		Oph	17 46 12.0 +05 43 00	4.2	III2p	70	best for binoculars
6475	7	Sco	17 53 50.0 −34 47 36	3.3	II2r	75	Ptolemy's Cluster, lowest Messier object
6611	16	Ser	18 18 45.0 −13 47 54	6.0	II3mn	8	in Eagle Nebula IC 4703
IC 4725	25	Sgr	18 31 45.0 −19 07 12	4.6	I2p	26	
IC 4756		Ser	18 38 54.0 +05 26 00	4.6	III2m	40	best for binoculars
6705	11	Sct	18 51 05.0 −06 16 12	5.8	I2r	11	Wild Duck Cluster, compact
7092	39	Cyg	21 31 52.0 +48 25 30	4.6	III2p	31	in crowed region
7654	52	Cas	23 24 48.0 +61 36 00	6.9	I2r	16	triangular shape
7789		Cas	23 57 28.6 +56 42 52	6.7	II2r	25	Caroline's cluster, rich and compact

Three star patterns (asterisms) should be mentioned. M 73 is a close ensemble of four stars (diameter about 1') in Aquarius, 1.3° east of the bright globular cluster M 72. Here a small telescope is required. Kemble's Cascade in Camelopardalis (04 00.0 +63 00) is a chain of 20 stars, extending over 2.5°, that ends at the open cluster NGC 1502. The asterism was presented by the Canadian amateur Lucian Kemble. The third example is the Coathanger, or Brocchi's Cluster, in Vulpecula (19 25.4 +20 11). Though it is cataloged by Collinder as an open cluster (Cr 399), it only is a random accumulation of five 6 mag stars, spread over 1°. Both patterns look best in binoculars (try to follow the cascade over some fields of view). However, there are many more asterisms. American amateurs are very creative in tracking down and naming such cases.

Kemble's Cascade, a remarkable 2.5° long chain of stars in Camelopardalis, ending at the open cluster NGC 1502 (lower felft).

Globular clusters are a worthwhile, albeit relatively small object group for different apertures and observation experiences. The brightest are ω Centauri and 47 Tucanae (NGC 104), appearing stellar for the naked eye. The latter stands only 2° northeast of the SMC but has nothing to do with our companion galaxy. Due to the fact that most globular clusters are concentrated around the centre of the Milky Way in Sagittarius, a considerable number are out of reach for northern observers. Fortunately, their hemisphere has its own showpieces, like M 13 or M 15, visible as tiny spots to the naked eye under dark sky conditions.

If you use binoculars or a small telescope with 6 cm aperture, all 29 globular clusters of the Messier catalogue as well as some brighter NGC objects can be observed. Alltogether, there are 115 NGC and 8 IC globulars. All, but six IC objects, were found visually. The observation with the smallest possible optics is certainly an interesting challenge. A 3-inch refractor is sufficient to detect single stars in the outer region of M 13 under suburban sky. For owners of medium-sized telescopes with apertures of 15 to 20 cm, a large number of globular clusters appear resolved, at least at the edge. In order to see the majority of the stars of a globular cluster of moderate concetration – getting the impression of a 'completely resolved' object – at least a 30 cm telescope is required.

Visual observers who have access to large telescopes can try the more unknown globular clusters. With a 50 cm telescope all 15 globular clusters of the Palomar catalogue are observable. In some of them, individual stars are already visible. But also, the observation of the brighter objects is very attractive with a large telescope. Often, they are field-filling at magnifications of 500-times. No details are visible, like star chains in M 15 running from the centre to far into the periphery.

The following table gives some interesting globular clusters. Very remote clusters are NGC 7006 and NGC 2419, with distances from the Sun of 134,000 and 269,000 light-years, respectively. With a distance of 7,200 light-years, M 4 is pretty close. Three objects do not belong to the Milky Way: M 54, NGC 1049, and G1. It is astonishing that a Messier object, M 54, is an extragalactic globular cluster. Lying at a distance of 86,400 light-years, it is an outer member of the Sagittarius dwarf galaxy, located at the opposite side of the Milky Way, although some have suggested it is actually the nucleus of that galaxy. The home galaxy of NGC 1049, about 460,000 light-years away, is the spheroidal dwarf system in Fornax which belongs to the Local Group. Though observing this galaxy is a real challenge, the globular cluster is already visible in a 10-inch telescope. The cluster G 1, shining at 13.7 mag, is the brightest globular cluster in the Andromeda Nebula. With a diameter of only 30", the 2.5 million light-years distant object, should be visible in a 12-inch at high magnification. The LMC offers 13 globular clusters, catalogued in the NGC; the brightest is NGC 1854.

G 1 (Mayall II) is the brightest globular cluster associated with the Andromeda Nebula. The tiny 13.5 mag object (diameter 30") is located 2.5° southwest of the M 31 centre.

The Shapley-Sawyer concentration class is an interesting quantity, influencing the visual appearance. It is a nice project to look-up globulars along the concentrations I to XII. However, one should select objects with comparable magnitudes or distances. The table offers two extremely compact cases (I): M 75 and NGC 7006. Extremely loose is NGC 5466 (XII) in Bootes, a very difficult object.

A little easier to find is NGC 5053 (XI) in the same constellation. It forms a seldom pair with another globular, M 53; the bright class V object is only 1° northeast. A wider pair is M 10 and M 12 in Ophiuchus. With +79°, M 53 and NGC 5053 have the highest galactic latitude of all globular clusters. NGC 288 in Sculptor stands opposite (–89°), almost at the southern pole of the Milky Way. The object is 1.7° southeast of the bright galaxy NGC 253. NGC 6544 in Sagittarius joins the galactic equator (–2°). Some globular clusters clearly show an oval appearance. M 19 in Ophiuchus is a fine example. M 71, located in the Milky Way of Sagitta, is a transition object between a globular and open cluster.

The table also contains two globular clusters, included in the famous Palomar list. The brighter one is Palomar 9, discovered earlier as NGC 6717. Though the object is easy to find, right between the 5 mag stars v^1 and v^2 Sgr, it is difficult due to its large size. The other case is Palomar 7 (IC 1276) in Serpens, a real challenge.

A rare globular cluster pair: M 53 and NGC 5053 in Coma Berenices. The distance between the compact Messier object (class V) and its loose companion (class XI) is 1°. The latter is a challenging object, a rich-field telescope is required. Both globulars are at the same distance of about 60,000 light-years.

Interesting globular clusters (m = visual magnitude, D = size in arc minutes)

NGC/IC	M	Con	Position	m	Class	D	Remarks
104		Tuc	00 24 05.2 −72 04 49	4.0	III	50	47 Tucanae, brightest globular, naked eye (stellar)
		And	00 32 46.5 +39 34 41	13.7	VIII	0.5	G 1 (Mayall II), brightest globular cluster in M 31
288		Scl	00 52 45.5 −26 35 51	8.1	X	13	lowest galactic latitude, 1.7° southeast of NGC 253
1049		For	02 39 48.1 −34 15 30	12.9	V	1.2	in Fornax System (Local Group)
1854		Dor	05 09 20.0 −38 50 51	10.4		2.3	brightest globular in LMC
1904	79	Lep	05 24 10.6 −24 31 25	7.7	V	9.6	
2419		Lyn	07 38 08.5 +38 52 57	10.3	II	4.6	Intergalactic Wanderer
5024	53	Com	13 12 55.3 +18 10 11	7.7	V	13	near north galactic pole
5053		Com	13 16 27.0 +17 41 55	9.0	XI	10	1° southeast of M 53
5139		Cen	13 26 47.0 −47 28 51	5.3	VIII	55	ω Centauri, largest globular, naked eye (stellar)
5272	3	CVn	13 42 11.2 +28 22 34	6.3	VI	18	some dark markings
5466		Boo	14 05 27.3 +28 32 04	9.0	XII	9.2	extremely loose
5904	5	Ser	15 18 33.8 +02 05 00	5.7	V	23	22' northwest of 5 Ser
6093	80	Sco	16 17 02.5 −22 58 28	7.3	II	10	bright stellar nucleus
6121	4	Sco	16 23 35.5 −26 31 29	5.4	IX	36	near Antares; among the nearest globulars
6205	13	Her	16 41 41.5 +36 27 39	5.8	V	20	curved dark features, near to galaxy NGC 6207
6218	12	Oph	16 47 14.5 −01 56 52	6.7	IX	12	pair with M 10, 3.3° southwest
6254	10	Oph	16 57 08.9 −04 05 56	6.6	VII	20	
6273	19	Oph	17 02 37.7 −26 16 05	6.8	VIII	5.3	oval shape
6341	92	Her	17 17 07.3 +43 08 13	6.5	IV	14	curved chains of stars
6544		Sgr	18 07 20.6 −24 59 49	7.5	V	8.4	nearly at the galactic equator
IC 1276		Ser	18 10 45.7 −07 12 40	10.3	XII	3.9	Palomar 7, very difficult
6656	22	Sgr	18 36 24.2 −23 54 10	5.2	VII	32	first discovered object
6715	54	Sgr	18 55 03.3 −30 28 43	8.4	III	9.1	member of Sagittarius Dwarf

6717		Sgr	18 55 06.2 −22 42 01	8.4	VII	21	Palomar 9, between v^1 and v^2 Sgr
6779	56	Lyr	19 16 35.5 +30 11 07	8.4	X	8.8	near M 57
6809	55	Sgr	19 39 59.4 −30 57 42	6.3	XI	19	
6838	71	Sge	19 53 46.1 +18 46 44	8.4	XI?	7.2	transition to open cluster
6864	75	Sgr	20 06 04.8 −21 55 15	8.6	I	4.5	very compact
6981	72	Aqr	20 53 27.9 −12 32 11	9.2	IX	6.6	M 73 (asterism) 1.3° east
7006		Del	21 01 29.5 +16 11 17	10.6	I	2.8	very compact
7078	15	Peg	21 29 58.3 +12 10 03	6.3	IV	18	contains the tiny planetary nebula Pease 1
7089	2	Aqr	21 33 27.2 −00 49 22	6.6	II	16	
7099	30	Cap	21 40 22.0 −23 10 43	6.9	V	12	23' west of 41 Cap

Diffuse Nebulae

The Messier catalogue contains 7 diffuse nebulae, 47 are listed in the NGC and 85 in the IC. One NGC object (the Maia Nebula NGC 1432) and all ICs were found by photography. The following table gives a selection of 43 diffuse nebulae, covering all relevant types. A magnitude is not listed. This is not a useful quantity for emission-line objects; the same applies in the case of reflection nebulae. 79 diffuse nebulae, catalogued in the NGC/IC, belong to the Magellanic Clouds. The most spectacular is the Tarantula Nebula (NGC 2070), a gigantic HII region in the LMC. Lacaille, using a tiny refractor, designated the object as 8 mag 'star' 30 Doradus. For many emission nebulae a filter is an effective tool, offering very different views. The specific type of filter to best observe the listed object, is mentioned in the table. Some large binoculars can be equipped with a filter and, using these, spectacular views of the North America Nebula (NGC 7000) or the California Nebula in Perseus (NGC 1499) are guaranteed.

Interesting diffuse nebulae (EN = emission nebula, RN = reflection nebula, BN = bipolar nebula, DN = dark nebula, SNR = supernova remnant); size in arc minutes.

NGC/IC	M	Con	Position	Type	Size	Filter	Remarks
281		Cas	00 52 53.8 +56 37 30	EN	35 × 30	UHC	
IC 59 IC 63		Cas	00 58 30.0 +31 25 00	EN	10 × 5 10 × 3	UHC	25' north-east of γ Cas
1333		Per	03 29 18.0 +61 00 00	RN	6 × 3		9 mag star on north side
1435		Tau	03 46 10.0 +23 45 54	RN	30 × 30		Merope Nebula, in Pleiades
1499		Per	04 01 10.0 +36 27 36	EN	120 × 60	Hβ	California Nebula, nearest HII region
1555		Tau	04 21 57.1 +19 32 07	RN	3 × 2		Hind's Variable Nebula, near T Tauri
IC 2118		Eri	05 04 54.0 −07 15 00	RN	180 × 60	UHC	Witch Head Nebula (= NGC 1909)
IC 405		Aur	05 16 29.4 +34 21 22	EN+RN	30 × 20	Hβ	Flaming Star Nebula
		Ori	05 27 30.0 +03 58 00	EN	600	OIII	Barnard's Loop, circular HII region
1952	1	Tau	05 34 31.9 +22 00 52	SNR	6 × 4	OIII	Crab Nebula (supernova 1054)
1976	42	Ori	05 35 17.1 −05 23 25	EN+RN	40 × 35		Orion Nebula, HII region, Trapezium (θ Ori) in centre
1977		Ori	05 35 18.0 −04 49 15	RN	20		with open cluster
1999		Ori	05 36 25.4 −06 42 57	RN	2 × 2		centre of large EN
2070		Dor	05 38 42.5 −69 06 03	EN	30 × 20		Tarantula Nebula, HII region in LMC
IC 434		Ori	05 41 00.0 −02 27 12	EN	60 × 10	Hβ	around B 33
		Ori	05 41 00.0 −02 27 30	DN	8 × 6		B 33, Horsehead Nebula
2024		Ori	05 41 42.0 −01 51 24	EN	30 × 30	UHC	Flame Nebula, HII region
2068	78	Ori	05 46 45.0 +00 04 48	RN	8 × 6		like a comet, three brighter stars involved
2163		Ori	06 07 49.5 +18 39 27	BN	3 × 2		cometary
IC 443		Gem	06 17 48.7 +22 49 04	SNR	50 × 40	UHC	
2237		Mon	06 30 54.6 +05 02 52	EN	80 × 50	OIII	Rosette Nebula, HII region, central cluster
2261		Mon	06 39 09.5 +08 44 40	BN	3 × 1		Hubble's Variable Nebula (cometary)
2264		Mon	06 40 58.2 +09 53 44	BN	20		Cone Nebula (cometary), in open cluster around 15 Mon

2736		Vel	09 00 17.0 −45 56 53	SNR	20 × 2	OIII	Herschel's Ray (Pencil Nebula)
3372		Car	10 45 06.0 −59 52 00	EN	120 × 120	Hβ	η Carinae Nebula, HII region, the star is embedded in the tiny Homunculus Nebula
		Crx	12 50 00.0 −62 30 00	DN	420 × 300		Coalsack, near α Crucis
		Oph	17 23 30.0 −23 38 00	DN	37 × 17		B 72, Snake Nebula
		Sgr	18 02 45.0 −27 49 54	DN	5 × 5		B 86, Ink Spot
6514	20	Sgr	18 03 26.0 −22 59 27	EN+DN	20 × 20	UHC	Trifid Nebula, HII region, open cluster, dark part (B 85)
6523	8	Sgr	18 03 42.0 −24 22 48	EN+DN	45 × 30	UHC	Lagoon Nebula, HII region, open cluster, dark parts (B 88, 89, 296), near M 21
		Sgr	18 15 32.0 −18 10 58	DN	7 × 7		B 92
6618	17	Sgr	18 20 47.0 −16 10 18	EN	20 × 15	OIII	Omega Nebula, HII region, open cluster
		Sct	18 39 11.0 −06 37 15	DN	10 × 10		B 103
6726 6727 6729		Cra	19 01 40.0 −36 53 00	RN+BN	80		NGC 6729 cometary, globular cluster NGC 6723 is 30' northwest
		Aql	19 06 05.0 −06 50 20	DN	4 × 4		B 133
		Aql	19 38 46.0 +10 28 53	DN	7 × 7		B 142/43
6888		Cyg	20 12 06.5 +38 21 18	EN	18 × 13	OIII	Crescent Nebula, due to strong stellar wind
		Cyg	20 27 27.1 +37 22 39	BN	3 × 3	Hβ	Sh2-106 (also designated S 106)
IC 1318		Cyg	20 22 14.0 +40 15 24	EN	50 × 30	UHC	γ Cygni Nebula, HII region
6992		Cyg	20 56 18.0 +31 44 30	SNR	60 × 8	OIII	Veil Nebula (western part)
6995		Cyg	20 57 10.0 +31 14 00	SNR	12 × 12	OIII	Veil Nebula (eastern part)
7000		Cyg	20 59 18.0 +44 31 00	EN	120 × 100	UHC	North America Nebula, HII region
IC 5146		Cyg	21 53 24.0 +47 16 00	EN+DN	10 × 10	Hβ	Cocoon Nebula, end of B 168
7635		Cas	23 20 45.0 +61 12 42	EN	15 × 8	OIII	Bubble Nebula, HII region

The two fanlike emission nebulae IC 59 and IC 63 are located about 25' north-east of the bright star γ Cassiopeiae.

The small bipolar nebula NGC 2163 in northern Orion looks like a comet.

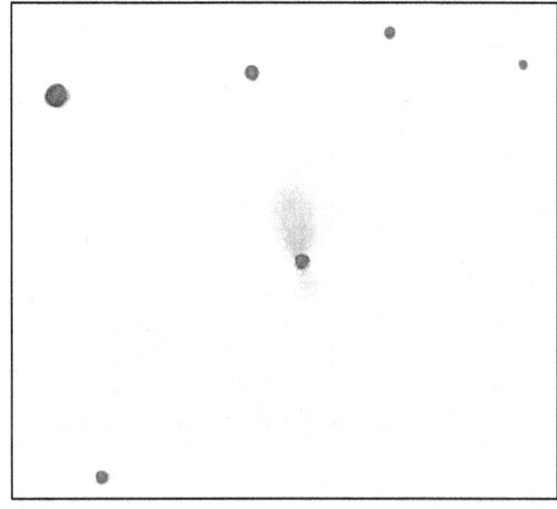

The listed supernova remnants (SNR) are emission-line objects, visually accessible with 8 inches aperture and OIII-filter. Old exemplars, shining in the Hα line, are not included; they are reserved for distinguished astrophotographers. The optimal visibility of a remnant is given when the filaments, like those in the Veil Nebula, can be resolved as such. If the aperture is too small or the feature too faint, the fine arc blurs to an area of low surface brightness. With increasing aperture, more and more details can be detected while the exit pupil remains the same, as the magnification increases and thus the apparent visual angle for our eye at constant surface brightness. In fact, it is important to choose exactly the right exit pupil for optimum detail perception.

The Crab Nebula M 1 is something special. Both the remnant and the central pulsar can be visually observed. Of course, the rotating neutron star can not be seen directly. What we perceive is the synchrotron radiation, i.e. the 'lighthouse' beam. It is the dominant whitish light of the nebula, which already Messier saw. The remnant itself is much more difficult to detect. It shows a filament-like structure, visible in a 14-inch telescope plus OIII-filter.

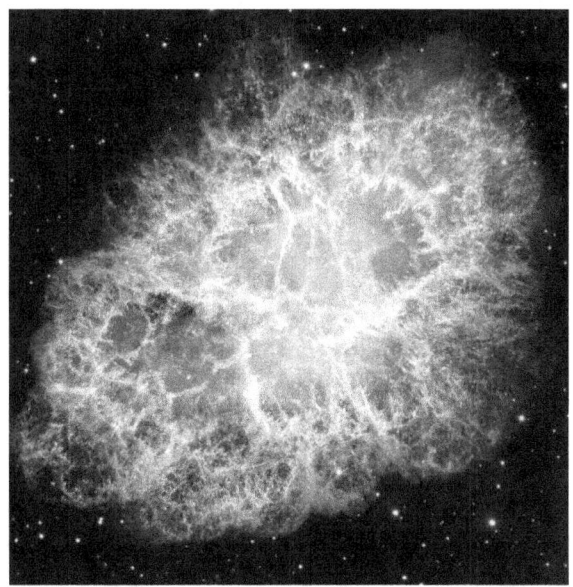

Visually, only the whitish synchrotron light of the Crab Nebula (M 1) can be seen, but with an OIII-filter a bit of the chaotic structure of the supernova remnant (as an emission nebula) appears, which is so dominant in astrophotographic images.

John Herschel discovered NGC 2736 in 1835. The curious nebula looks like a pencil. It is a small optically visible part of the large Vela supernova remnant (emitting X-rays). Since the distance is only 815 light-years, it must have been an extremely bright event (about −11 mag). It took place about 11,000 years ago. The remnant contains a pulsar.

Herschel's Ray NGC 2736 is part of the large Vela supernova remnant. The strange object is also known as 'Pencil Nebula'.

Compared to emission line objects, the visual observation of reflection nebulae poses a different challenge for the amateur. Due to the continuous spectrum, filters do not help and a dark sky is even more important than with other objects. A good example is the Merope Nebula (NGC 1435) in the Pleiades. While the object is very difficult with larger aperture, a 2-inch can show it. However, the glare of the star is a problem. Try to keep stray light out of the field of view.

The reflection nebula around the double star is NGC 6726/27. The small cometary nebula south-east is NGC 6729 with the variable star R CrA is its head.

Attractive reflection nebulae are the Witch Head Nebula (IC 2118) in Eridanus or M 78 and NGC 1999 in Orion. A nice double object is NGC 6726/27 in Corona Australis. Only 12' southwest is the small fan-shaped nebula NGC 6729, originating from the variable star R CrA. This nice trio was visually discovered by Julius Schmidt in 1861 at Athens Observatory.

Some dark nebulae can be seen with the naked eye. The best-known example is the Coalsack in Crux. It was considered by the Australien aborigines as part of the 'dark constellation' the Emu with α Crucis as its eye. Northern observers enjoy the Great Rift, a dark area that divides the band of the Milky Way in Cygnus – the largest dark nebula in the sky. In binoculars it disperses into countless dark structures. This is where one of the biggest difficulties in dark nebula observation comes into play. The sky can offer completely different views depending on the optics used. It often remains unclear, which object is a dark nebula or merely a field without stars. In binoculars, for example, dark fields can be noticed that are invisible with a large telescope – but the latter, in turn, shows a small black area which is invisible in binoculars. A good example is B 86 in Sagittarius; Edward Emerson Barnard coined the name 'Ink Spot'.

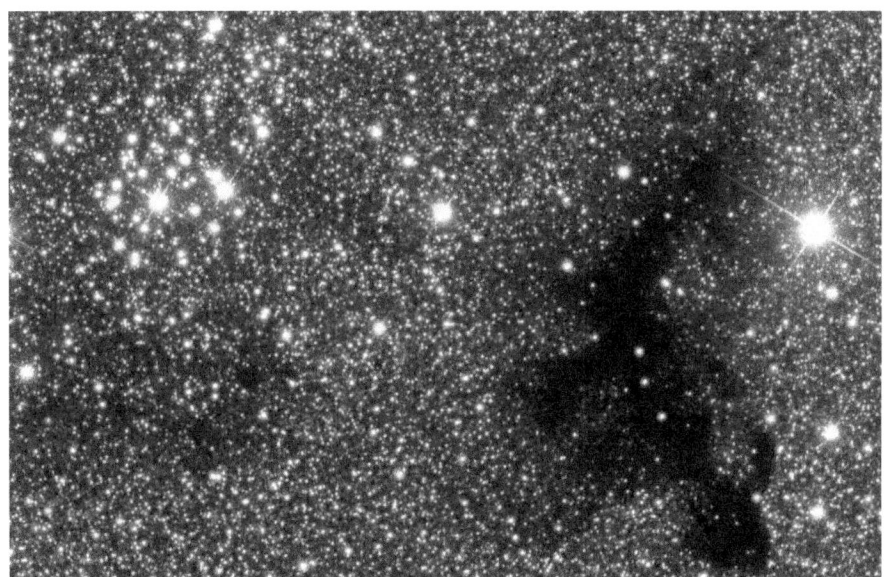

The small dark nebula B 86, known as the Ink Spot, is located near the open cluster NGC 6520 in Sagittarius.

The situation is quite different with the few objects that stand out against the background of a bright nebula. The prime example of this is the Horsehead Nebula (B 33) in Orion, lying in front of IC 434. But we also have the cometary shaped Cone Nebula in NGC 2264, the dark bars in the Trifid Nebula, and the dark nebulae in the Lagoon Nebula. The most striking of the numerous dark structures around η Carinae is the Homunculus Nebula. Actually, it is deep red, but the eye is not sensitive to this colour, thus it appears black in the telescope.

About 200 dark nebulae are suitable for visual observation. As a beginner, it is best to first orient yourself to light objects. Later one can consult Barnard's great catalogue. Many of its entries are beautiful targets for visual observers, such as B 142/43 in Aquila or the Snake Nebula (B 72) in Ophiuchus. However, caution is necessary, because the Barnard collection is based on photographic images taken in the 1920s. Many objects are therefore barely visible by visual means.

Planetary Nebulae

About 500 planetary nebulae can be observed with a 20-inch telescope. The Messier catalogue containes 4; 94 are in the NGC and 36 in the IC. Two IC objects (IC 972, IC 1454) are in the famous list, compiled by George Abell (don't confuse them with his objects from the catalogue of rich galaxy clusters). The following table gives a selection of interesting objects. A few of these can already be seen with binoculars; the brightest is M 27 (7.4 mag). Only a few dozen planetaries are larger than 5'; many do not exceed 40". A difficult case is the Helix Nebula NGC 7293 in Aquarius. Though the integrated brightness is about 7th mag, the diameter is 11'. An extremely small example is IC 2501 in Carina.

The observational techniques are generally different for faint, large objects and small, bright ones. In order to optimally capture a large planetary like NGC 7293 or NGC 1360, the lowest possible magnification should be used. The maximum exit pupil is achieved (depending on the aperture) at a minimum magnification. Then the nebula appears as bright as for the naked eye. The reason why the observer subjectively sees it easier, however, is magnification: the larger the luminous area, the brighter it is perceived by the eye.

Things are different for small, bright objects. Their shape is often defined so well that a high magnification is best. A good example is NGC 7026 in Cygnus. However, stellar objects can only be detected with an OIII filter, best used in a filter wheel or slider. When one quickly observes the field alternately with and without the filter, the planetary will flash out. While the stars appear normal without the filter, the filter dims them – except the emission object. A similar result is provided by using a visual spectroscope. Actually, IC 2501 in Carina was discovered by Williamina Fleming in 1904 with such a device. While stars, emitting a continuous spectrum, are displayed as small colour bands, the nebula image remains stellar, because usually only one emission line is present.

Interesting planetary nebulae (m = visual magnitude, D = size in arc minutes, CS = magnitude of the central star)

NGC/IC	M	Con	Position	m	D	CS	Remarks
40		Cep	00 13 01.0 +72 31 21	12.3	1.2	10.6	C 2, most northern NGC planetary
246		Cet	00 47 03.3 −11 52 17	10.9		11.9	near NGC 247, lowest galactic latitude
650/51	76	Per	01 42 18.1 +51 34 17	10.1	3.1	17.0	Little Dumbbell, like M 27
1360		For	03 33 14.6 −25 52 16	9.4	6.4	11.4	very difficult
1514		Tau	04 09 17.0 +30 46 35	10.9	2.2	9.4	bright central star
1535		Eri	04 14 15.8 −12 44 20	9.6	0.9	12.2	light blue
2242		Aur	06 34 07.4 +44 46 40	14.7	0.4	17.6	faintest NGC planetary
2392		Gem	07 29 10.8 +20 54 45	9.1	0.9	10.5	Eskimo Nebula, light green
2438		Pup	07 41 50.6 −14 44 05	10.8	1.3	17.7	in open cluster M 46, light green
2818		Pyx	09 16 10.0 −36 37 34	11.2	1.4		in open cluster Mel 96
IC 2501		Car	09 38 47.3 −60 05 29	10.7	0.03	14	stellar
3242		Hya	10 24 46.1 −18 38 31	7.7	1.1	12.0	Ghost of Jupiter, greenish
3587	97	UMa	11 14 47.7 +55 01 10	9.9	2.8	15.9	Owl Nebula, high galactic latitude, 48' southeast of galaxy M 108
3918		Cen	11 50 17.8 −57 10 55	8.1	0.4	10.8	Blue Planetary, looks like Uranus
4361		Crv	12 24 30.8 −18 47 03	10.9	2.1	13.2	central star easy visible, high galactic latitude
IC 972		Vir	14 04 26.0 −17 13 39	13.9	0.9	17.7	Abell 37
6210		Her	16 44 29.5 +23 48 02	8.8	0.4	12.9	blue-green
6302		Sco	17 13 44.1 −37 06 12	9.6	1.5		Bug Nebula, bright western knot
6543		Dra	17 58 33.4 +66 38 01	8.1	0.3	11.4	Cat's Eye Nebula, knot IC 4677, near pole of ecliptic
6720	57	Lyr	18 53 35.1 +33 01 47	8.8	3.0	14.8	Ring Nebula, central star very difficult, SC galaxy IC 1296 4' north-west
6826		Cyg	19 44 48.2 +50 31 32	8.8	0.6	10.4	Blinking Planetary, greenish
6853	27	Vul	19 59 36.3 +22 43 18	7.4	6.7	13.9	Dumbbell Nebula, easiest planetary
7008		Cyg	21 00 32.8 +54 32 38	10.7	1.4	12.3	

		Cyg	21 02 18.8 +36 41 38	14.0	0.5		Egg Nebula, bipolar protoplanetary nebula
7009		Aqr	21 04 10.8 −11 21 47	8.0	0.6	11.5	Saturn Nebula, light green
7026		Cyg	21 06 18.6 +47 51 10	10.9	0.4	14.3	elongated, blue-green
7027		Cyg	21 07 01.7 +42 14 12	8.5	0.3	16.3	bluish
IC 5150		Gru	21 59 35.1 −39 23 06	11.0	2.2		= IC 5148, annular
IC 1454		Cep	22 42 24.5 +80 26 35	14.0	0.6	18.0	Abell 81
7293		Aqr	22 29 38.4 −20 50 11	7.3	17.6	13.5	Helix Nebula, very large, UHC filter useful
7662		And	23 25 53.9 +42 32 08	8.3	0.6	13.2	Blue Snowball

The celebrated Ring Nebula M 57 in Lyra has a much less prominent 'companion': the 14 mag Sc galaxy IC 1296, located only 4' north-west. Try a 14-inch telescope to see the 234 million light-years distant object.

The disc-like 10.8 mag planetary nebula NGC 3918 in Centaurus has a strong blue color. It was discovered by John Herschel in 1834 and looks much like the planet Uranus.

The 10.7 mag planetray IC 2501 in Carina is almost stellar. It was discovered by visual spectroscopy.

The spectroscope can also be used to discover faint central stars. Often, the surface brightness of the nebula is so high that the central star is drowned out by it, although it would have been nominally visible. In the spectroscope, the fine continuous spectrum of the star stands out in the nebula. Planetaries to which this technique can be successfully applied are, for instance, the Cat's Eye Nebula (NGC 6543) in Draco and NGC 6210 in Hercules. In a few cases the central star is easily seen: NGC 40, NGC 2392, NGC 6543, NGC 7009. A special case is NGC 6826 in Cygnus, called the Blinking Planetary. When looking directly, the bright central star stands out, but the nebula disappears. Using indirect (averted) vision, one perceives the surrounding nebula, but the star disappears. Another curious case should be mentioned: NGC 1514 in Taurus (William Herschel's 'star with an atmosphere'). There is a 9.4 mag star inside the round nebula. Curiously, it is not the physical central star. The true one – and originator of the nebula – is the hidden, very close companion.

Some planetaries are associated with an open or globular cluster. From stellar evolution theory, this sounds plausible. The most prominent case is NGC 2438 in M 46 (Puppis). A 4-inch telescope is enough to see the small, round nebula inside the bright open cluster – a wonderful sight. However, NGC 2438 is in fact a line of sight object and is not part of the cluster. NGC 2818 lies in the open cluster Mel 96 (Pyxis), but is also not thought to be a member of the cluster (sometimes called NGC 2818A). A more interesting example is the case of Pease 1 in the globular cluster M 15 (Pegasus). Just finding this tiny, faint object (diameter 1", 14 mag) in the dense stellar environment of the globular cluster is a great challenge.

Galaxies, Quasars, and Galaxy Clusters

Galaxies are worthwhile objects for viewing with every aperture and magnification. If one zooms into a bright barred spiral, new details are always revealed. With a 4-inch you barely see an elongated nebula. With 8 inches you will notice a slight structure with dust stripes. Spiral arms and a bar appear with 12 inches aperture and a 16-inch will show branches and knots. Faint galaxies can be seen with a large telescope like bright ones in a small one, so you start again from the beginning. However, the use of filters – valuable tools for emission line objects – is limited to broadband filters that merely block artificial light.

Interesting galaxies, sorted by RA (m = visual magnitude, size in arc minutes, PA in °).

NGC/IC	M	Con	Position	m	Type	Size	PA	Remarks
55		Scl	00 15 08.0 −39 13 10	7.9	SBm	31.2 × 5.9	108	edge-on, asymmetric
IC 10		Cas	00 20 24.5 +59 17 33	10.4	IBm	6.4 × 3.5	135	in zone of avoidance, Local Group
205	110	And	00 40 22.1 +41 41 07	8.1	E5 pec	19.5 × 11.5	170	M 31 companion, dark feature
221	32	And	00 42 41.8 +40 51 57	8.1	E2	8.5 × 6.5	179	M 31 companion, compact
224	31	And	00 42 44.3 +41 16 08	3.4	Sb	189 × 61	35	Andromeda Nebula, contains super star cluster NGC 206
253		Scl	00 47 33.1 −25 17 15	7.2	SBc	29.0 × 6.8	52	Sculptor Galaxy, Silver Dollar, edge-on, dark lane
292		Tuc	00 52 38.0 −72 48 00	2.3	SBm	319 × 205	45	Small Magellanic Cloud, Local Group
404		And	01 09 26.9 +35 43 06	10.3	E0	3.5 × 3.5		Mirach's Ghost, near β And
598	33	Tri	01 33 51.9 +30 39 29	5.7	Sc	68.7 × 41.6	23	Triangulum Nebula, bright HII region NGC 604
628	74	Psc	01 36 41.7 +15 47 00	9.4	Sc	10.5 × 9.5	25	face-on, 'grand design'
2573		Oct	01 41 53.2 −89 20 03	13.4	Sb	1.9 × 0.7	85	Polarissima Australis
891		And	02 22 33.0 +42 20 50	9.9	Sb	11.7 × 1.6	22	edge-on, dark lane
1365		For	03 33 36.7 −36 08 27	9.6	SBb	11.0 × 6.2	32	face-on, Fornax cluster
		Dor	05 23 35.0 −69 45 20		SBm	600 × 540		Large Magellanic Cloud, Local Group
2403		Cam	07 36 50.6 +65 36 06	8.5	SBc	23.4 × 11.8	127	contains super star cluster NGC 2404
2903		Leo	09 32 09.7 +21 29 57	9.0	SBc	12.6 × 6.0	17	contains HII region NGC 2905

NGC	M	Con	RA/Dec	Mag	Type	Size	PA	Notes
3031	81	UMa	09 55 33.5 +69 04 02	6.9	Sb	24.9 × 11.5	157	Bode's Nebulae, pair with M 82
3115		Sex	10 05 14.1 −07 43 05	8.9	S0	7.2 × 2.4	40	Spindle Galaxy
3314		Hya	10 27 12.8 −27 40 00	13.1	Sa/Sab	2.1 × 1.5	143	optically superimposed galaxies
3351	95	Leo	10 43 57.8 +11 42 12	9.7	SBb	7.4 × 5.0	13	pair with M 96
3368	96	Leo	10 46 45.8 +11 49 12	9.3	SBab	7.8 × 5.2	176	
3379	105	Leo	10 47 49.5 +12 34 52	9.3	E1	5.3 × 4.8	71	trio with NGC 3384/89
3556	108	UMa	11 11 29.4 +55 40 22	10.0	Sc	8.6 × 2.4	80	edge-on, star superimposed, 48' northwest of planetary nebula M 97
3172		UMi	11 47 14.5 +89 05 37	14.4	Sb	1.0 × 0.7	39	Polarissima Borealis
4254	99	Com	12 18 49.3 +14 25 03	9.9	Sc	5.3 × 4.6	51	Virgo Cluster
4258	106	CVn	12 18 57.8 +47 18 25	8.4	SBbc	18.6 × 7.2	150	dark features
4374	84	Vir	12 25 03.6 +12 53 13	9.1	E1	6.5 × 5.6	135	Markarian Chain, Virgo Cluster
4406	86	Vir	12 26 11.5 +12 56 47	8.9	E3	8.9 × 5.8	130	Markarian Chain, Virgo Cluster
4565		Com	12 36 20.5 +25 59 16	9.6	Sb	15.8 × 2.1	136	edge-on, very flat
4594	104	Vir	12 39 59.3 −11 37 21	8.0	Sa	8.6 × 4.2	89	Sombrero Galaxy, edge-on, dark lane, not in Virgo Cluster
4826	64	Com	12 56 43.8 +21 40 59	8.5	Sab	10 × 5.4	115	Black Eye Galaxy, dark structure
4889		Com	13 00 08.3 +27 58 35	11.5	E3	2.8 × 2.0	80	cD in Coma cluster
5236	83	Hya	13 37 00.2 −29 52 02	7.5	Sc	12.9 × 11.5	44	face-on, 'grand design'
5457	101	UMa	14 03 12.4 +54 20 58	7.9	Sc	28.8 × 26.9	26	Pinwheel Galaxy, face-on, bright HII regions
5866	102	Dra	15 06 29.4 +55 45 49	9.9	S0-a	6.5 × 3.1	128	edge-on, dark lane
5907		Dra	15 15 53.8 +56 19 49	10.3	Sc	12.6 × 1.4	155	edge-on, very flat
6822		Sgr	19 44 56.6 −14 48 23	8.7	IBm	15.4 × 14.2	5	Barnard's Galaxy, Local Group
6946		Cyg	20 34 52.1 +60 09 12	8.8	SBc	11.5 × 9.8	57	in zone of avoidance
7331		Peg	22 37 05.1 +34 25 13	9.5	Sbc	10.2 × 4.2	171	near Stephan's Quintet
7479		Peg	23 04 56.7 +12 19 20	10.9	SBc	4.0 × 3.1	25	face-on barred spiral

All 40 galaxies of the Messier catalogue are visible in a 4-inch, as well as about 1,200 NGC objects brighter than 12 mag. The *New General Catalogue* offers 6,022 galaxies and the *Index Catalogue* another 3,972, however, some are about 16 mag. An 18-inch telescope should show them all under optimal conditions. But this only means detection. A deeper study needs much larger apertures. There are amateurs observing with 48-inch reflectors – and still can not see all the structures, visible on photographs, made with much smaller instruments. The following table lists 40 interesting 'normal' galaxies for various apertures (peculiar objects are presented in a separate table).

The brightest and largest galaxies are naked eye objects: the Magellanic Clouds and the Andromeda Nebula (M 31). Under dark sky conditions (see the Bortle scale), M 33 in Triangulum can be added. Objects for binoculars are M 81 or M 101 in Ursa Major. In a larger telescope, you have the problem that they hardly fit in the field of view, even at lowest power.

Areas outside the Milky Way are, of course, predestined for galaxy observation. Extreme places are the celestial poles. The galaxy nearest to the northern one is NGC 3172 (Polarissima Borealis), at a distance of 1°. The opposite place is occupied by NGC 2573 (Polarissima Australis), at a distance of ¾°. Both objects should be visibile with a 10-inch. The observation of the pole regions is easy with a Dobsonian, but problematic with an equatorial monting. Another special place is Hubble's zone of avoidance. Only a few brighter objects are visible here. The crowded fields and extinction by dust in the Milky Way make them difficult to observe. Examples are IC 10 in Cassiopeia or NGC 6946 in Cepheus, with galactic latitudes of 3° and 11°, respectively.

NGC 6946 in Cepheus is located in a crowded field of the Milky Way.

Sometimes a galaxy is superimposed by a star. If it appears dominant, this can cause confusion. The star could be thought of as the compact centre or even a supernova. Examples are M 108 in Ursa Major and NGC 6207 in Hercules; the superimposed stars are pretty central, shining at 12 and 13.5 mag, respectively. Actually, there were some erroneous supernova recordings in such cases.

The 14.4 mag spiral galaxy UGC 2885 in Perseus is superimposed by a 10 mag star.

Less trouble offers the nice celestial meeting of a galaxy and a bright star. The classic example is Mirach's Ghost, NGC 404, only 6.8' northwest of β And. The bright star is problematic: it outshines the galaxy and kills your dark adaptation. Try to keep it out of the field. This leads to a general rule: don't relax your observing session with a short visit to Jupiter or Saturn. Herschel, while sweeping, occasionally got these planets into the field of view. This brought him an unwanted half hour break.

The edge-on galaxy NGC 5907 in Draco (north to the right).

Very popular amongst observers are edge-on galaxies, i.e. spiral systems with high inclination (i > 80°). If the disc is very thin and the bulge small, it is defined to be a flat galaxy. The essential parameter is the axial ratio (a/b). Of course, not every edge-on galaxy is flat. Consider the prominent Sombero Galaxy (M 104) in Virgo. Due to 84° inclination, we see it edge-on, but the object is not flat. The dominant bulge causes an axial ratio of 2. The issue is different for NGC 5907. The Sa galaxy has i = 87° and is really flat: a/b = 9.

Superthin galaxies, like NGC 100 in Pisces or IC 2233 in Lynx, can have axial ratios up to 16:1. In some cases, the disc is warped, for example like the strange Integral Sign Galaxy (UGC 3697) in Camelopardalis. Visually these objects are very difficult to see. The problem is to perceive a narrow line of low surface brightness. The trick is to use indirect (averted) vision and field sweeping. For the latter, swing the telescope perpendicular to the line (its orientation in the field of view should be known). At least 14 inches aperture is needed to perceive a superthin galaxy. However, to see the integral sign, you need about 20 inches.

The superthin Integral Sign Galaxy UGC 3697 in Camelopardalis has a warped disc, seen edge-on. The strange object forms a pair the peculiar spiral UGC 3714, about 8' south-east.

The following table shows a sample of galaxies with abnormal features. Most are due to gravitational interaction of two or more objects. Close encounters cause strong tidal forces which deform the galaxies. A prominent example is M 51 and its companion NGC 5195. Another interesting case is M 82 in Ursa Major. The galaxy is strongly disturbed, but there is no obvious cause. Actually, this is M 81, located 37' south and looking uninvolved. M 87, the central galaxy in the Virgo Cluster, is famous for its optical jet, emanating from the central supermassive black hole. For the 20" long feature, photographically discovered by Heber Curtis in 1918, at least a 20-inch reflector and high magnification are required. NGC 5128, the nearest radio galaxy (Centaurus A), was first seen as a collision of an E- and S0-galaxy. We now believe that the chaotic dark lane comes from strong internal activity. The evidence is due to the detection of a long radio jet and strong infrared radiation, a sign for violent star formation. Other starburst galaxies are NGC 4449 in Canes Venatici, showing an amorphous shape, and NGC 6240 in Ophiuchus, one of the most luminous infrared sources.

Galaxies with peculiar features (m = visual magnitude, size in arc minutes, PA in °).

NGC/IC	M	Con	Position	m	Type	Size	PA	Remarks
100		Psc	00 24 02.6 +16 29 11	12.7	Scd	6.2 × 0.6	55	superthin galaxy
		Scl	00 37 41.1 −33 42 59	14.5	Ring	1.1 × 0.9		Cartwheel, ring galaxy
660		Ari	01 43 02.0 +13 38 39	12.0	Ring	8.3 × 3.1	170	polar ring galaxy
985		Cet	02 34 37.4 −08 47 06	13.5	Ring	1.0 × 0.9	69	Seyfert galaxy, ring galaxy
1068	77	Cet	02 42 40.8 −00 00 46	8.9	Sb pec	7.1 × 6.0	70	Seyfert galaxy
1275		Per	03 19 48.1 +41 30 41	11.9	S0 pec	2.3 × 1.6	110	Seyfert galaxy, Perseus cluster
		Cam	07 11 19.2 +71 50 12	12.9	Sd pec	3.2 × 0.3	90	Integral Sign Galaxy, warped disc
IC 2233		Lyn	08 13 59.5 +45 44 23	12.3	SBd	5.2 × 0.6	173	superthin galaxy, pair with NGC 2537
2623		Cnc	08 43 24.1 +25 45 17	13.4	Sb pec	2.4 × 0.7	60	merging system
2685		UMa	08 55 34.9 +58 44 05	11.5	SB0 pec	4.6 × 2.5	38	Helix Galaxy, polar ring galaxy
3034	82	UMa	09 55 54.0 +69 40 59	8.4	Sd pec	11.2 × 4.3	65	Bode's Nebulae, pair with M 81
4449		CVn	12 28 11.3 +44 05 42	9.6	IBm	6.2 × 4.4	45	starburst galaxy, bright HII regions
4486	87	Vir	12 30 49.4 +12 23 26	8.6	E2 pec	8.3 × 6.6	170	Virgo Cluster, jet

4650A		Cen	12 44 49.0 −40 44 49	13.3	Ring	1.6 × 0.8		polar ring galaxy, in Centaurus Chain
4861		CVn	12 59 01.8 +34 51 43	13.5	SBm pec	4.2 × 1.6		cometary galaxy
5023		CVn	13 12 11.8 +44 02 14	12.1	Scd	7.3 × 0.8	28	superthin galaxy
5128		Cen	13 25 29.0 −43 00 58	6.8	S0 pec	25.7 × 20.0	35	Centaurus A, chaotic dark lane
5194	51	CVn	13 29 52.6 +47 11 44	8.4	Sbc	11.2 × 6.9	7	Whirlpool Nebula, face-on, pair with NGC 5195
		Ser	15 17 14.4 +21 35 08	16.0	Ring	0.3 × 0.3		Hoag's Object, ring galaxy
6166		Her	16 28 38.5 +39 33 05	11.8	cD pec	2.2 × 1.5	32	multiple nucleus
6240		Oph	16 52 58.8 +02 24 11	12.9	E pec	2.1 × 1.0	20	starburst galaxy
6745		Lyr	19 01 41.9 +40 45 33	12.5	Sm pec	1.3 × 0.5	24	merging system
7252		Aqr	22 40 44.8 −24 40 42	11.4	SB0 pec	2.1 × 1.7	119	merging system

The superthin galaxy IC 2233 in Lynx is only 17' apart from the the peculiar Bear Paw Galaxy NGC 2537 (north to the right).

Other peculiar galaxies are NGC 2623 in Cancer, NGC 6745 in Lyra, and NGC 7252 in Aquarius (here a merger of two galaxies takes place). NGC 6166, the cD galaxy in the rich cluster Abell 2199, has a multiple nucleus, the result of several mergers in the dense environment. Even stranger are ring galaxies. Examples are the Cartwheel in Sculptor, Hoag's Object in Serpens, and NGC 985 in Cetus. A special case is called 'polar ring galaxy'. The prototype is NGC 4650A, a member of the Centaurus Chain. All these objects are visually very difficult. Large apertures and high magnifications are needed to see a bit of the peculiar structures. To photograph small galaxies, a long focal length is recommended.

The climax of extragalactic activity is seen in quasars. Though their optical appearance is rather inconspicuous, they bear many mysteries. The following table shows some interesting 'normal' quasars (QSO) and BL Lacertae objects (BL). The fascination has two reasons: the extreme distance, like in the case of S5 0014+81 in Cepheus, and the idea that a supermassive black can be watched at work. For a long time, the 12.8 mag quasar 3C 273 in Virgo was the only exemplar occasionally visited by amateurs. This is due to a lack of information. With 8 inches aperture, about 50 objects should be observable, to a 12-inch about 100. Decisive for the visual success, especially with very faint quasars, is a good finder chart. The rest is star hopping (or GoTo). Observation requires high magnification to reduce the background noise. Very faint quasars appear in perception during indirect vision and disappear again. As always, honesty is required: even a negative result is valuable because many objects are variable. Thus, the question about the brightest quasar is rethorical. It is confirmed that some, like W Com, can be brighter than 12 mag. BL Lac varies between 12.4 and 17 mag.

The stellar object BL Lac is the namesake of a special class of quasars. An aperture of 14 inches is needed to see the 900 million light-years distant variable object.

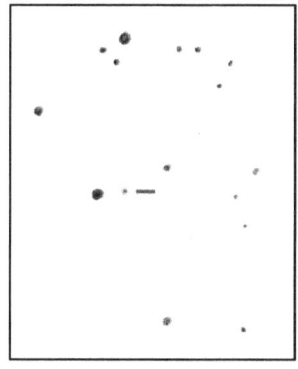

The bright quasar 3C 273 in Virgo is 2.1 billion light-years away. The jet, due to synchrotron radiation, emerges from the central supermassive black hole.

Interesting quasars (BL = BL Lacertae object, m = visual magnitude, distance in billion light-years)

Object	Type	Con	Position	m	Dist	Remarks
S5 0014+81	QSO	Cep	00 17 08.1 +81 35 07	15.2	22.2	extreme distance
I Zw 1	QSO	Psc	00 53 34.9 +12 41 36	14.0	0.80	faint halo
3C 48	QSO	Tri	01 37 41.3 +33 09 35	16.2	4.65	variable
3C 66A	QSO	And	02 22 39.6 +43 02 08	15.2	5.53	variable
Q 0957+561	QSO	UMa	10 01 20.9 +55 53 52	16.7	13.8	Double Quasar, gravitational lens
Mrk 421	BL	UMa	11 04 27.2 +38 12 32	12.9	0.41	variable
W Com	BL	Com	12 21 31.7 +28 13 58	11.5	1.38	variable
Mrk 205	QSO	Dra	12 21 44.2 +75 18 39	15.2	0.96	optical pair with NGC 4319
3C 273	QSO	Vir	12 29 06.7 +02 03 08	12.8	2.11	variable, jet
Mrk 501	BL	Her	16 53 52.2 +39 45 36	13.4	0.47	faint halo
BL Lac	BL	Lac	22 02 43.3 +42 16 39	12.4	0.95	variable

There is an interesting duo, located in Draco. The partners are the quasar Mrk 205 and the galaxy NGC 4319. The former is only 42" south of the barred spiral. Though the redshifts are very different, Halton Arp claimed a physical connection. This would imply that redshift is not a measure of distance. His evidence came from a 'bridge of light', detected between both objects on long-exposed plates. Later this was rejected due to high-resolution digital images. Thus, we have an optical pair and the Hubble-Lemaître law still holds true.

The controversial case of the quasar Mrk 205 (marked) in the 'halo' of the nearby galaxy NGC 4319 in Draco (the other galaxy is NGC 4291)

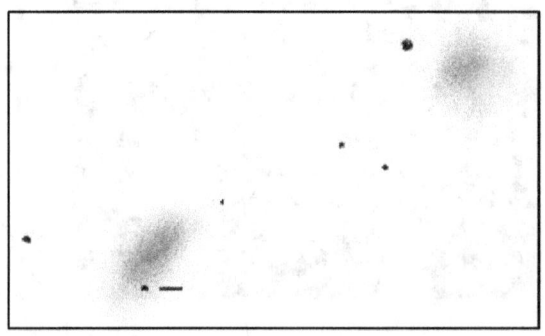

One object is exceptional: Q 0957+561, the Double Quasar in Ursa Major. It is located 14' north of the edge-on galaxy NGC 3079 and consists of two stellar objects, lying 6" apart. However, there is only one physical source: the double image is the result of gravitational lensing due to a faint galaxy lying between us and the quasar (see HST image). With indirect vision, you can glimpse the common light of 16.7 mag in a 14-inch telescope. But to separate the 'pair', at least 20 inches, a high magnification and the best seeing is required.

The Double Quasar Q 0957+561 in Ursa Major, north of the edge-on galaxy NGC 3079 (in trio with NGC 3073, bottom right, and PGC 28990); inset: HST image.

There are many nice galaxy pairs and trios in the sky; the following table shows some examples. Take for instance NGC 7332/39 in Pegasus; here two bright edge-on galaxies with different orientiations are 5' apart. The brighter one, NGC 7332, has a boxy bulge. In some cases, interaction is visible.

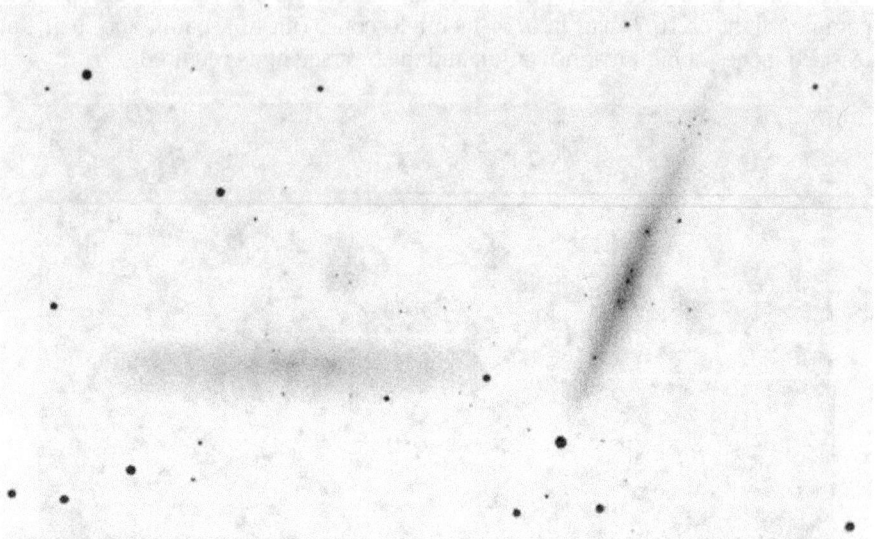

The stunning pair NGC 7332/39 in Pegasus: two edge-on galaxies in different orientation (drawing by Wilhelm Tempel)

There are close pairs, where tidal forces have let to deformations. NGC 4485/90 in Canes Venatici is a duo of a spiral and an irregular galaxy, pretty near to β CVn. The interaction is less obvious for NGC 5544/45 in Bootes, a very close pair of an edge-on and a face-on galaxy, looking like an exclamation mark. The Antennas (NGC 4038/39) in Corvus show longs tidal tails; the nuclei are only 0.9' apart. Similar is the case of The Mice (NGC 4676) in Coma Berenices, supposedly playing with each other.

Though a very close pair too, the Siamese Twins (NGC 4567/68) in Virgo are only a chance alignment. Obviously, NGC 750/51 in Triangulum is not an optical pair; the two E0 galaxies are in contact and have a common envelope. It is quite difficult to resolve the double galaxy. No interaction is visible for NGC 4298 and NGC 4302 in Coma Berenices, but they are in fact a physical pair (both are members of the Virgo Cluster). The different inclinations of the two Sc galaxies give an impressive view.

NGC 4567 and NGC 4568, the Siamese Twins in Coma Berenices – an optical pair.

A sample of galaxy pairs and trios, with at least two companions within 15 arc minutes distance (D); C = components.

NGC	C	Con	Position	m	Size	PA	Type	D	Remarks
1	a	Peg	00 07 15.9 +27 42 32	12.9	1.7 × 1.2	120	Sb	ab 1.9	optical pair
2	b	Peg	00 07 17.1 +27 40 43	14.2	1.0 × 0.6	112	Sab		optical pair
467	a	Psc	01 19 10.1 +03 18 05	11.8	1.7 × 1.7		S0	ab 10.9	in background
470	b	Psc	01 19 44.8 +03 24 33	11.8	2.9 × 1.7	155	Sb	bc 5.8	
474	c	Psc	01 20 06.7 +03 24 31	12.0	7.1 × 6.3	75	S0	ac 15.8	
750	a	Tri	01 57 32.7 +33 12 32	12.2	1.6 × 1.3		E0 pec	ab 0.4	dumbbell, in contact
751	b	Tri	01 57 33.0 +33 12 10	11.5	1.2 × 1.2		E0 pec		dumbbell, in contact
1618	a	Eri	04 36 06.6 −03 08 55	12.7	2.4 × 0.8	26	SBb	ab 13.2	edge-on, near ν Eri
1622	b	Eri	04 36 36.6 −03 11 18	12.5	3.7 × 0.7	33	SBab	bc 10.9	edge-on, near ν Eri

			RA/Dec	Mag	Size	PA	Type	Sep	Notes
1625	c	Eri	04 37 06.3 −03 18 14	12.3	2.1 × 0.5	131	SBb		edge-on, near ν Eri
M 65	a	Leo	11 18 55.6 +13 05 27	9.3	9.8 × 2.9	174	Sa		Leo Triplet
M 66	b	Leo	11 20 15.1 +12 59 24	8.9	9.1 × 4.1	173	Sb		Leo Triplet
3628	c	Leo	11 20 16.7 +13 35 24	9.5	13.1 × 3.1	104	Sb		Leo Triplet, edge-on, dark lane
4038	a	Crv	12 01 52.8 −18 51 52	10.3	3.4 × 1.7	94	SBm	ab 0.9	The Antennae
4039	b	Crv	12 01 53.8 −18 53 08	10.4	3.3 × 1.7	55	SBm		The Antennae
4206	a	Vir	12 15 16.7 +13 01 22	12.0	6.4 × 1.1	0	Sbc	ab 11.3	edge-on, Virgo Cluster
4216	b	Vir	12 15 54.0 +13 08 52	10.3	8.1 × 1.8	19	SBb	bc 11.4	edge-on, Virgo Cluster
4222	c	Com	12 16 22.6 +13 18 25	13.2	3.1 × 0.5	56	Scd	ac 24.0	edge-on, Virgo Cluster, in background
4298	a	Com	12 21 32.9 +14 36 24	11.4	3.2 × 1.9	140	Sc	ab 2.4	Virgo Cluster
4302	b	Com	12 21 42.2 +14 35 54	11.9	5.3 × 1.0	178	Sc		edge-on, Virgo Cluster
4485	a	CVn	12 30 31.3 +41 42 03	11.9	2.4 × 1.8	15	IBm pec	ab 3.5	near β CVn
4490	b	CVn	12 30 36.1 +41 38 34	9.8	6.4 × 3.2	115	SBcd		near β CVn
4567	a	Vir	12 36 32.7 +11 15 28	11.3	3.1 × 2.2	85	Sbc	ab 1.4	Siamese Twins, optical pair
4568	b	Vir	12 36 34.2 +11 14 19	10.8	4.6 × 2.2	23	Sbc		Siamese Twins, optical pair
4627	a	CVn	12 41 59.6 +32 34 26	12.4	1.7 × 1.0	26	E4 pec	ab 2.5	
4631	b	CVn	12 42 07.6 +32 32 30	9.2	15.2 × 2.8	86	SBcd		Whale Galaxy, edge-on
4656	a	CVn	12 43 58.1 +32 12 33	10.5	15.3 × 2.4	33	SBm pec	ab 0.3	edge-on
4657	b	CVn	12 44 06.9 +32 12 33	12.4	1.3 × 0.6	160	I pec		
4676A	a	Com	12 46 10.1 +30 43 57	13.5	1.4 × 0.6	0	SB0-a	ab 0.5	The Mice
4676B	b	Com	12 46 11.2 +30 43 21	13.8	2.2 × 0.8	2	S0-a		The Mice
5544	a	Boo	14 17 02.4 +36 34 16	13.3	1.1 × 1.0		SB0-a	ab 0.4	in contact
5545	b	Boo	14 17 04.8 +36 34 29	15.0	1.0 × 0.3	58	Sbc		in contact
7332	a	Peg	22 37 24.6 +23 47 53	11.1	4.1 × 1.1	155	S0 pec	ab 5.2	edge-on, boxy
7339	b	Peg	22 37 47.0 +23 47 11	12.2	2.8 × 0.7	93	SBbc		edge-on, different orientation

NGC 4627/31 and NGC 4656/57 are two neighboring pairs in Canes Venatici. Each is made of a large elongated galaxy (NGC 4631, NGC 4656) and a small, close companion (NGC 4627, NGC 4657). With a slight movement of the telescope of about 30' you go from one pair to the other. Also impressive are trios like the popular Leo Triplet or NGC 467/70/74 in Pisces. Another visually fascinating case is located in Eridanus: NGC 1618, NGC 1622, NGC 1625. The trio which form a 18' long chain of edge-on galaxies with different orientations, lying 11' northeast of the 4th mag star ν Eri. We find a similar trio in Virgo: NGC 4206, NGC 4216 and NGC 4222; here the chain is 23' long (no star).

NGC 4627/31 and NGC 4656/57, two neighboring pairs in Canes Venatici.

The next step is galaxy groups; the following table presents some interesting examples, many from the Hickson catalogue. An easy one is the Leo Quartet. More difficult is the group HCG 61, also known as The Box (Coma Berenices). The four members look very different, only three are actually physically associated. The famous Stephan's Quintet in Pegasus is a group of five galaxies 30' southwest of the large galaxy NGC 7331 in Pegasus. However, the brightest galaxy of the quintet, NGC 7320, is much nearer to us and thus not a physical member of the group. Thus, we actually have a quartet. But there is the fainter galaxy NGC 7320C, associated with the other four, so we get a quintet again. The sequence continues with groups of six and seven galaxies (Seyfert Sextet, Copeland Septet).

A sample of galaxy groups (N = number of members, D = group diameter in arc minutes, m = range of visual magnitudes)

Group	Con	Position	N	D	m	Brightest members
NGC 383 group	Psc	01 07.4 +32 24	14	20	12.2–16.5	NGC 373, 375, 379, 380, 382–88
NGC 507 group	Psc	01 33.4 +03 20	20	30	11.3–16.0	NGC 494–6, 498, 501, 503, 504, 507, 508, IC 1684, 1685, 1688–90
Leo Quartet	Leo	10 18.0 +21 49	4	20	10.8–12.9	NGC 3185, 3187, 3190, 3193
Copeland Septet	Leo	11 37.8 +21 59	7	6.5	13.7–15.2	NGC 3745, 3746, 3748, 3750, 3751, 3753, 3754
The Box	Com	12 12.4 +29 11	4	8	12.2–13.5	NGC 4169, 4173, 4174, 4175
Markarian Chain	Vir	12 28.0 +13 10	9	180	8.9–11.8	M 84, M 86, NGC 4435, 4438, 4458, 4461, 4473, 4477, 4479; Virgo Cluster
Centaurus Chain	Cen	12 44.3 –40 44	5	30	11.8–14.7	NGC 4622, 4622B, 4650, 4650A (polar ring)
Seyfert Sextet	Ser	15 59.2 +20 45	5	2.5	13.2–15.4	NGC 6027
Stephan's Quintet	Peg	22 36.0 +33 58	5	6	12.5–13.6	NGC 7317, 7318 A/B, 7319, 7320 (not a member), 7320C

The Markarian Chain in the centre of the Virgo Cluster, starting with M 84 and M 86 (bottom left: M 87).

The appearance of galaxy groups ranges from spherical arragements to chains. One of the best known is the Markarian Chain in the central region of the Virgo Cluster, spanning over 1.5° from M 84/86 to NGC 4477. The best southern chain is located in Centaurus, which is, however, three magnitudes fainter. It contains the spectacular polar ring galaxy NGC 4650A. As an intermediate step between galaxy groups and clusters, there are larger ensembles containing several dozen galaxies, such as those around NGC 383 or NGC 507 in Pisces. The brightest members should be visible with a telescope of 8 inches aperture. It is always a special experience to see many galaxies at once (with a sufficient field of view).

Galaxy clusters are huge aggregates, containing galaxies of all types; the following table lists some bright examples. The nearest is the Virgo Cluster, which is the dominant part of the Local Supercluster (followed by the Fornax cluster). Of course, the visual observation of a galaxy cluster is limited to the brightest members, usually located in the central region (an exception is the irregular Hercules cluster). An 8-inch telescope can already show some of them. Following Abell, the brightness of a galaxy cluster is defined by the magnitude of the 10th brightest member (m_{10}).

A sample of bright galaxy clusters (m_{10} = magnitude of the 10th brightest galaxy, D = diameter in arc minutes, Dist = diastance im million light-years, SC = supercluster membership)

Abell	Name	Con	Position	m_{10}	D	Dist	SC	Main member(s)
262		And	01 52.8 +36 08	13.3	100	196	Per-Psc	NGC 708
347		And	02 25.8 +41 52	13.3	56	253	Per-Psc	NGC 910
426	Perseus	Per	03 18.6 +41 30	12.5	190	237	Per-Psc	NGC 1275
1060	Hydra	Hya	10 36.9 −27 30	12.7	168	174	Hya-Cen	NGC 3309/11
1367	Leo	Leo	11 44.5 +19 50	13.5	100	285	Com/A1367	NGC 3842
	Virgo	Vir	12 27.0 +12 43	10.0	480	54	Local SC	M 87
1656	Coma	Com	12 59.8 +27 58	13.5	220	336	Com/A1367	NGC 4874/89
2151	Hercules	Her	16 05.2 +17 44	13.8	50	504	Her	NGC 6042
4449	Fornax	For	03 38.5 −35 27	10.3	180	58	Local SC	NGC 1399

Also interesting are Abell 262 in Pisces (elongated), Abell 347 in Andromeda (located south of the edge-on galaxy NGC 891), Abell 426 in Perseus (hosting the turbulent galaxy NGC 1275), and Abell 2151 in Hercules (showing very different galaxy types). We also have Abell 1367 in Leo and Abell 1656 in Coma Berenices, the dominant members of the Coma/A1367 Supercluster. On the sky Abell 1656 and Abell 1367 are 20° apart. Abell 1656 contains two cD galaxies: NGC 4874 (11.9 mag) and NGC 4889 (11.5 mag) which are easily visible with a telescope of 8 inches aperture. With superclusters we have reached the endpoint of the cosmic hierarchy. Conclusion: visually or by photography, we are able to follow all steps of the overall picture.

The rich cluster of galaxies Abell 426 in Perseus (note the chain-like structure). The brightest member is the peculiar S0 galaxy NGC 1275 (above centre), also known as the radio source Perseus A.

Appendix

Bibliography

Archinal B., Hynes S., Star Clusters, Willmann-Bell 2003

Bracken C., The Deep-sky Imaging Primer, Deep-sky Publishing 2017

Bratton M., The Complete Guide to the Herschel Objects, Cambridge University Press 2011

Burnham R. jr., Burnham's Celestial Handbook, 3 Vol., Dover Publublications 1978

Buta R., Corwin H., Odewahn S., The de Vaucouleurs Atlas of Galaxies, Cambridge University Press 2007

Dick S. J., Discovery and Classification in Astronomy, Cambridge University Press 2013

Ferris T., Galaxies, Stewart, Tabori & Chang Publications 1982

Glyn Jones K., Messier's Nebulae and Star Clusters, Cambridge University Press 1991

Hoskin M., The Construction of the Heavens: William Herschel's Cosmology, Cambridge University Press 2012

Houston W. S., Deep Sky Wonders, Sky Publishing Corporation 1999

Hynes S., Planetary Nebulae, Willmann-Bell 1992

Kanipe J., Webb D., Arp Atlas of Peculiar Galaxies, Willmann-Bell 2006

Kepple G. R., Sanner G., The Night Sky Observers Guide, 2 Vol., Willmann-Bell 1998

König M., Binnewies S., The Cambridge Atlas of Galaxies, Cambridge University Press 2017

Mollan C. (ed.), William Parsons, 3rd Earl of Rosse: Astronomy and the Castle in Nineteenth-Century Ireland, Manchester University Press 2016

O'Meara S., Herschel 400 Observing Guide, Sky Publishing Corporation 2007

O'Meara S., Hidden Treasures, Sky Publishing Corporation 2007

O'Meara S., Messier Objekts, Sky Publishing Corporation 1998

Roy J.-R., Unveiling Galaxies, Cambridge University Press 2017

Sandage A., Bedke J., The Carnegie Atlas of Galaxies, Carnegie Institution of Washington 1994

Steinicke W., Jakiel R., Galaxies and how to observe them, Springer 2006

Steinicke W., Observing and Cataloguing Nebulae and Star Clusters – from Herschel to Dreyer's New General Catalogue, Cambridge University Press 2010

Stoyan R., Atlas of the Messier Objects, Cambridge University Press 2008

Stoyan R., interstellarum Deep Sky Guide, Cambridge University Press 2018

Streicher M., Astronomical Delights, BSB Printers Plokwane/Petersburg 2012

Webb Society Deep Sky Observer's Handbook, 7 Vol., Enslow Publishers 1979-87

Willasch D., Slootegraaf A., Pearls of the Southern Skies, Firefly Books 2014

General Index

Bold numbers refer to figures. Frequent terms, like nebula, cluster, or galaxy (in their various forms), are omitted.

100-inch reflector (Hooker)	20, 26, 62
18¼-inch reflector	**15**
18.7-inch reflector	**12**
200-inch reflector (Hale)	62, 67, 95
21 cm line	50, 59, 71
36-inch reflector	19, 20, 21
40 foot reflector	14
72-inch reflector	14, 19, 20, 21, 22, 24
Abell catalogue	54
Abell, George	36, 37, 38, 76, 116, 135
absolute brightness	32, 33, 42, 60, 69, 70, 71
absorption line	24
accelerated expansion	78
accretion disc	60, 73
active galactic nucleus (AGN)	38, 72, 73, 74
air pollution	89, 95
air refractor	45
air turbulence	88
Alcmene	4
Almagest	4
Al-Sufi	5, 39
altazimuth mounting	11, 31, 86, 87, 97
Alter, Jiří	36
Ambartsumian, Victor	44
American Astronomical League	86
amorphous	32
angular momentum	40, 79
annular nebula	55, 118
Antoniadi scale	89
Antoniadi, Eugène-Michel	89
aperture	11, 14, 18, 86, 87, 95, 96, 127, 135
apex (solar)	59
apparent brightness	32, 33, 61
apparent size	33, 61, 70
Aratus	4
Archinal, Brent	36
Aristotle	4
Armagh Observatory	16, 21
Arp, Halton	38, 67, 128
association	30, 44
asterism	30, 44, 45, 104, 109
asteroid	93
astrograph	26
asymmetric	32
Athens Observatory	115
Atlas (titan)	39
Atlas Coelestis	9
Atlas of Peculiar Galaxies	38, 67
atmospheric condition	2, 88
axial ratio	124
azimuth	31
Bahcall, John	38
bar	57, 59, 64
Barnard, Edward Emerson	37, 115, 116
baryonic matter	77
Bath	9, 17
Bayer letter	31
Bevis, John	6, 7
Big Bang	72, 77, 78, 81, 82
big galaxy	26
binoculars	1, 45, 91, 103, 104, 115
bipolar	52
Birr Castle	16, 19, 20, **21**, 22, 23, 24
BL Lacertae object	30, 73, 74, 127, 128
black hole (stellar)	42, 50, 79
black hole (supermassive)	57, 60, 72, 73, 74, 79, 80, 82, 125, 127
Blazar	73
blind spot	92
blue globular	48
blue straggler	46
blueshift	27, 80
Bode, Johann Elert	8, 45
bolometric brightness	33
Book of the Fixed Stars	5
Bortle scale	90, 91
Bortle, John	90
bottom-up model	77
boxy	63, 66, 67, 130, 132
British Catalogue	9, 13
broadband filter	88, 120
bulge	46, 57, 59, 60, 64, 65, 71, 124, 130
bulge-to-disc ratio	65
Bunsen, Robert	25

Caldwell Catalogue	34	cosmic background radiation	78, 82
Cape catalogue	15	cosmic distance ladder	69
Cape Town	6, 15	cosmological constant	77
carbon	52	cosmological standard model	77
Catalog of Bright Nebulae (LBN)	36	Crossley reflector	96
		Curtis, Heber	26, 125
Catalog of Dark Nebulae (LDN)	37	Cysat, Johann Baptist	5
		dark adaption	90, 93, 95, 123
Catalog of Galaxies and Galaxies (CGCG)	38	dark cloud	4, 48, 57
		dark energy	77, 78
Catalog of Open Cluster Data	36	dark frame	98
Catalog of Principal Galaxies (PGC)	38	dark lane	56, 120, 121, 125, 126, 132
Catalog of Star Clusters and Associations	36	dark matter	72, 76, 78, 79, 82
		Darquier, Antoine	53
Catalogue of Galactic Planetary Nebulae (PK)	37	d'Arrest, Heinrich	94, 96
		Datchet	9, 11
CCD camera	95, 97	de Vaucouleurs, Gérard	62
cD galaxy	64, 67, 71, 81, 121, 126, 136	density	40, 46
		density wave	79
Cederblad, Stefan	36	dew	90, 95
celestial equator	30, 31	differential rotation	40, 59
celestial pole	31, 34, 90, 101, 122	diffuse nebula	1, 8, 30, 36, 49, 54, 101, 109
central star	54, 55, 117, 119		
centrifugal force	27	digital camera	2, 55, 96, 97
Cepheid	26, 69, 70	Digitized Sky Survey (DSS)	85
Chéseaux, Jean-Phillipe Loys de	6, 8	direct vision	92, 95, 119
		disc	40, 57, 65
Church of Ireland	22	distance	28
classification	60, 62	Dobson, John	86
cold dark matter (CDM)	77	Dobsonian	86, 87, 90, 93, 95
Collinder, Per	36, 39	double star	9, 11, 89, 96, 103, 114
comet	11, 20, 49, 53, 101		
comet seeker	86	Doppler Effect	27, 41, 59, 70, 72
cometary	49, 111	Doppler, Christian	27
cometary galaxy	68	Draper, Henry	25, 26
compact galaxy	61	drawing	94, 95
compact group of galaxies (HCG)	38	Dreyer, John Louis Emil	**16**, 21, 24, 34
		DSLR camera	96, 98
concentration class	47	dumbbell	52, 55, 68, 131
cones	92	Dunlop, James	15
constellation	1, 8, 32, 44, 45, 47, 67, 85, 90, 105	dust cloud	51, 53, 59, 60
		dwarf elliptical	64
continuous spectrum	24, 25, 50, 114, 116, 119	dwarf spheroidal	64
		ecliptic	30
contrast	2, 87, 88, 93	ecliptical coordinates	31
convection	89	edge-on	61, 65, 120, 121, 124, 130, 131, 132, 135
cooling	97		
Copeland, Ralph	24, 25		
Corwin, Harold	38, 76		

Einstein, Albert	27, 70, 72, 77	Fraunhofer, Joseph	24
elevation	31	front view (reflector)	12, 14
elongation	61, 62	fusion (nuclear)	40, 41
entrance pupil	93	GAIA satellite	56
epoch	31	galacic length	37
equatorial coordinates	30	galactic centre	53, 60
equatorial mouting	86, 87, 97	galactic coordinates	31, 37, 38
equinox	31	galactic latitude	37, 53, 56, 117, 122
ESO/Uppsala Survey	35	galactic (nebula)	30, 49
event horizon	73	galactic plane	50, 53
excitation	49	galactic year	41
exit pupil	**87**, 93, 95, 113, 116	galaxías	4
expansion (universe)	72, 76	galaxy (term)	22
exposure time	96, 97	Galilei, Galileo	39
extended object	32, 33, 96	gaseous nebula	49
extinction	52	Gegenschein	90
extragalactic (nebulae)	17, 30, 31, 49, 55, 56, 127	General Catalogue (GC)	16
		General Relativity	70, 72, 77
extragalactic globula cluster	48	George III	9, 14
extragalactic object	30	Glasgow Observatory	22
extreme halo globular	47	globular cluster (name)	10
Eye & Telescope (software)	85	GoTo	91, 127
eyepiece	2, 87, 88, 95	grand design (galaxy)	6, 61, 120, 121
Faber-Jackson relation	69, 70	gravitational lens	128, 129
face-on	61, 65, 92, 120, 121, 126, 130	great debate	26
		Green, David	37
faintest star (fst)	90	Greenwich	6
Feldhausen	15	guest star	50
field of view	2, 11, 12, 87, 88, 93, 94, 95, 114, 124, 135	Guide (software)	85
		Gum, Colin	36
field sweeping	92, 95, 124	Halley, Edmond	6, 15, 45
filament	28, 78, 79, 113	halo	57, 72, 73, 128
filter	2, 95, 109, 110, 114	Harris, William	36
filter wheel	116	helium	40, 41, 42, 46, 50, 54, 56, 78
finderscope	91		
Fisher-Tully relation	69, 70	helium flash	41
fixed star	58	Henry brothers	34
Flamsteed number	31	Hera	4
Flamsteed stars	13	Heracles	4
Flamsteed, John	6, 13	Herbig-Haro object	36
flat galaxy	124	Herschel 2500	86
flatfield	98	Herschel 400	86
Fleming, Williamina	116	Herschel catalogues	13, 14, 86, 100
focal length	11, 14, 87, 95, 96, 98	Herschel class	13
		Herschel, Caroline	**9**, 11, 12, 13, 15
focus	97	Herschel, John	**9**, 13, 14, 15, 16, 19, 20, 23, 94, 95, 96, 100, 113, 119
Foucault, Léon	25		
fragmentation	40, 77		

Herschel, William	1, **9**, 10, 11, 12, 13, 14, 17, 18, 19, 20, 21, 22, 25, 34, 49, 53, 55, 66, 85, 93, 100, 119, 123	integrated brightness	33, 61, 116
		interacting galaxies	37, 94
		interferometry	60
		interstellar matter/medium	2, 39, 40, 41, 51, 52, 54, 56, 71, 74, 81
Hertzsprung-Russell Diagramm (HRD)	42	ionisation	49
Hevelius, Johannes	45	iris	93
HI region	30, 50	iron	41
Hickson, Paul	38	irregular (galaxy)	49, 66, 67
hierarchical structure	1, 2, 28, 79	Isfahan	5
HII region	30, 36, 49, 50, 54, 61, 68, 69, 81, 88, 109, 110, 111, 120, 121, 125	island universes	26
		ISO value	98
		Jansky, Karl	60
		jet	50, 73, 125, 127, 128
high velocity cloud	80		
Himmelsscheibe von Nebra	39	Kant, Immanuel	17
Hind, John Russell	51	Kant-Laplace theory	17, 19
Hipparchus	1, 4	Keeler, James	96
Hodierna, Giovanni, Battista	5, 8, 39, 45, 48, 101	Kensington	21
Holmberg, Eric	66	Kepler rotation	71
horizontal system	31	Kirchhof, Robert	25
hot pixel	98	Koehler, Johann Gottfried	8
Hubble Atlas of Galaxies	62	Kohoutek, Luboš	37
Hubble Deep Field (HDF)	79, **80**	Lacaille, Nicolas-Louis de	6, 15, 109
Hubble flow	70	ΛCDM model	77
Hubble parameter	27, 70	Laplace, Pierre-Simon de	17
Hubble sequence	65, 66	large sweeper	11
Hubble Space Telescope (HST)	46, 69, 79, 98	Lassell, William	94
		Legentil, Guillaume	6
Hubble type	62, 64, 66, 67	lenticular	63, 65, 67
Hubble, Edwin	1, 26, 27, 56, 62, 69, 79, 122	Leviathan of Parsonstown	20, 21, 24
		Lick Observatory	96
Hubble-Humason redshifts	27	light pollution	**89**, 90, 95
Hubble-Lemaître law	27, 28, 70, 77, 128	line filter	88
Huggins, William	25	line spectrum	24, 25
Humason, Milton	27	lithium	78
hydrogen	25, 40, 42, 46, 49, 50, 54, 56, 71, 78	lobe	52
		Lord Rosse	1, 7, 14, 19, **20**, 21, 22, 23, 24, 26, 94, 95, 96
hydrogen molecule	52		
Hynes, Steven	36	Lowell Observatory	26
Hα (line)	49, 55, 97, 113	LRGB image	98
Hα (filter)	88	lucky imaging	98
Hβ (line)	50, 54	luminance	97
Hβ (filter)	88, 110, 111	luminosity	33, 41, 42, 60, 69, 70, 71
Ihle, Abraham	6, 45		
inclination	50, 61, 65, 72, 124, 130	Lund Observatory	36
Index Catalogue (IC)	17, 122	Lynds, Beverly	36, 37
indirect (averted) vision	90, 92, 95, 119, 124	Lyngå, Gosta	36
infrared	56, 125		

Lyon Groups of Galaxies (LGG)	38	OIII (line)	50, 54
		OIII (filter)	88, 110, 111, 113, 116
Lyon-Meudon Extragalactic Database (LEDA)	38	Olowin, Ronald	38, 76
Magalhães, Fernão de	5	off/on-axis	97
Magellanic Clouds	64	opacity	52
magnification	2, 12, 87, 93, 95, 96, 106, 101, 113, 116, 127	orientation	61, 130, 132
		Oxford	16
		oxygen	25, 50, 54
main sequence	42	Palomar Observatory Sky Survey (POSS)	36, 37, 38, 85, 95
Marius, Simon	5		
mass	41, 60, 71	pancake	77
mass-luminosity ratio	71	panetarium program	31
Méchain, Pierre	8	parallax	69
megaparsec	69	Paranal Observatory	70
Melotte, Philibert Jacques	36, 39	Paris Observatoy	34
Messier Catalogue	10, 11, 14, 33, 34, 53, 100, 109, 122	parsec	32, 70
		Parsons, Laurence	16, 21, 23
Messier marathon	86	Parsons, William	19, 20
Messier, Charles	6, 7, 8, 10, 45, 53, 113	Paturel, Georges	38
		peculiar	67
metals	40, 56	Peebles, James	77, 79
Milky Way	18, 22, 64	Peiresc, Nicolas Clude Fabri de	5
Millenium Simulation	**78**		
monochromatic	92, 97	Peiresc, Nicolas-Claude Fabri de	48
Moore, Patrick	34		
Morphological Catalog of Galaxies (MCG)	38	perception	93, 113
		Perek, Luboš	37
Mount Palomar	62, 67, 72	periodic system	24
Mount Wilson	20, 26, 47, 62, 76	period-luminosity relation	26, 69
moving group	30, 44	Philosophical Transactions of the Royal Society	13, 15, 21
narrow-band filter	96		
NASA Extragalactic Database (NED)	36, 85	Photographic Atlas of Selected Regions of the Milky Way	37
natural history	17		
nebular hypothesis	19, 21, 23, 25	photographic magnitude	33
nebulous matter	10, 17, 19, 23, 25	photometry	26
neutron star	41, 50	Pickering, William	25
New General Catalogue (NGC)	16, 34, 122	piggyback	98
		pixel	95, 98
Newton, Isaac	17, 89	planet	49, 81, 119
Newtonian	9, 11, 12, 86	planetarium program	85, 100
NGC/IC	24, 33, 34, 35, 66, 67, 85, 86, 109	planetary nebula (name)	53
		planetary system	40
Nichol, John Pringle	22, 23	point source	32, 33, 96
Nilson, Peter	38	polar distance	13
nitrogen	25, 50	polar ring galaxy	35, 68, 125, 126, 127, 134, 135
object-specific catalogue	30, 85		
oblate	62, 63	polar sequence	90
observing site	2, 95, 96	pole of ecliptic	117

pole of Milky Way	107	Schmidt, Julius	115	
population (I, II)	40, 46, 57, 59	Schmidt, Maarten	72	
position angle (PA)	61, 120, 125	Schmidt-Cassegrain telescope (SCT)	31, **83**, 86, 87	
precession	31			
pretty pictures	94, 97	Scientific Papers of Sir William Herschel (Dreyer)	16	
prolate	62, 63			
proper distance	72	scintillation	89	
proper motion	31, 41, 58, 70	Secchi, Angelo	94	
protogalaxy	79, 82	seeing	88, 89, 95, 96, 98	
protoplanetary nebula	54, 118	sensitivity	98	
protostar	40	Seyfert, Carl	72	
Ptolemy	4, 39	Seyfert galaxy	72, 125	
pulsar	50, 113	Shakhbazian, Romela	75	
PushTo	91	Shapley, Harlow	26, 39, 47	
quasar	1, 38, 71, 72, 73, 74, 94, 100, 120, 127, 128, 129	Shapley-Sawyer class	48, 106	
		Sharpless, Stewart	36	
		signal-to-noise ration	97	
radial velocity	27, 41, 58	SIMBAD Astronomical Database	36, 85	
radiation pressure	40			
radio source	60, 67	Simeis objects	36	
radio telescope	50, 59	singularity	73	
radio wave	56, 60	sky coverage	13	
raw image	98	sky darkness	88	
Rayleigh scattering	50	sky motion	95	
red dwarf	60	sky quality meter (SQM)	91	
red giant	41, 42, 50	Slipher, Vesto	26	
redshift	27, 70, 74, 128	Slough	14, 15, 20	
refractor	86, 97	Sloan Digital Sky Survey (SDSS)	28, 70, 85	
relative position	13, 31, 32			
resolvability	24	Slough catalogue	15, 21	
Revised New General and Index Catalogue	35	small sweeper	11	
		Soneira, Ray	38	
RGB filters	97	Solar system	1, 9, 17, 82	
rich-field telescope	86	South, James	21, 23	
rigid rotation	59	space motion	31, 41, 44, 58	
ring galaxy	68, 125, 126	space view	87	
Roberts, Isaac	26	spectral class/type	36, 42, 50	
Robinson, Thomas Romney	21, 22, 23	spectrograph/-scope	24, 25, 70, 116, 119	
rods	92, 93	speculum	10, 20	
rotational velocity	59, 60, 70, 71	speed of light	28, 72, 76	
Royal Astronomical Society	13	spherical	66	
Royal Society	13, 21	spiral arms/structure	40, 57	
RR Lyrae star	47, 69	spiral nebula	19, 26	
Ruprecht, Jaroslav	36	spiral arms/structure	7, 22, 23, 24, 40, 44, 55, 57, 58, 59, 61, 64, 65, 71, 79, 92, 94, 120	
San Francisco	86			
Sandage, Allan	62			
saturation	97	St Helena	6	
Sawyer, Helen	47	standard colours	97	
Schmidt camera	35, 95	star chain	102	

star cloud	8, 34, 35, 57, 81	Trouvelot, Étienne	95
star formation	49, 52, 66, 79, 125	Trümpler classification	43, 44
star group	30	Trümpler, Robert	36, 39, 43
star hopping	127	tuning fork	**62**, 65
star review	9	tunnel view	87
star stream parallax	41	UBV system	33
star trails	97	UHC filter	88, 110, 111, 118
starburst (galaxy)	71, 125, 126	ultra-luminous infrared galaxy (ULIRG)	71
starhopping	91		
Steinicke, Wolfgang	2, 94, appendix	ultraviolet	33, 49, 54, 60
stellar evolution	42, 119	universe	1, 18, 24, 26, 28, 60, 66, 70, 72, 73, 76, 77, 78, 79, 80, 84
stellar wind	40, 52		
Stock, Jürgen	36		
stratum	18	Uppsala General Catalog (UGC)	38
stray light	90, 114	uranium	41
submillimeter wave	56	van den Bergh, Sydney	37
super star cluster	48, 120	variable star	10, 51, 53, 69, 73, 114, 115
supercluster	2, 28, 30, 38, 76, 82, 135		
		velocity-distance relation	27
supergiant	41, 42	vernal equinox	30
superimposed	121, 123	Véron, Phillipe	38
supernova	41, 50, 69, 77, 81, 110, 123	Véron-Cetty, Marie-Paule	38
		Vespucci, Amerigo	5
supernova remnant (SNR)	7, 18, 30, 37, 42, 49, 54, 110	viewing angle	87
		vignetting	93, 98
superthin galaxy	65, 125, 126	Virgo Consortium	78
surface brightness	33, 61, 88, 91, 93, 113, 119, 124	Virgo flow	70
		VISTA telescope	70
sweep	12, 13, 123	visual magnitude	33, 108, 117, 120, 125
synchrotron radiation	50, 73, 113, 127		
Table Mountain	15	void	28, 78, 79, 82
telephoto lens	98	Vorontsov-Velyaminov, Boris	37, 38
Telrad	91		
Tempel, Wilhelm	34, 94, 95, 96, 130	warped disc	124, 125
temperature	40, 42, 78, 90, 95	Whipple, John Adams	26
The Distribution of Rich Clusters of Galaxies (Abell)	38	white dwarf	41, 42, 69
		white nebulae	25
thermal noise	97	wind	84, 89, 95
TheSky (software)	85	Windsor	9, 13
Third Cambridge Catalogue of Radio Sources (3C)	72	working list	15
		X-ray	60, 113
tidal force	80, 125	Young, Charles	25
tidal tail	74, 94	Zeldovich, Yakov	77, 79
top-down model	77	zenith	90
tracking	85, 87, 93, 96, 97, 98	Zeus	4
		Zone Catalogue	13, 14
transparency	88, 95	zone of avoidance	56, 76, 120, 121
triaxial	62, 63	Zwicky, Fritz	38, 76
trigonometric parallax	41		

Object Index

Bold numbers refer to figures, italics to tables. Frequent objects, like Milky Way, or sun, are omitted. Greek letters are listed as words (α = Alpha).

11 Mon	*103*
12 Mon	6
15 Mon	*103*, *111*
30 Dor	109
3C 48	72, *128*
3C 66A	*128*
3C 273	72, 73, 127, **127**, *128*
41 Cap	109
47 Tucanae	46, 105, *108*
5 Ser	*108*
52 Cyg	**99**
Abell 37 (PN)	37, *117*
Abell 76 (PN)	68
Abell 81 (PN)	37, **37**, *118*
Abell 262	135, *135*
Abell 347	135, *135*
Abell 426	135, *135*, **136**
Abell 1060	*135*
Abell 1367	76, *135*, 136
Abell 1656	76, *135*, 136
Abell 2029	71
Abell 2151	135, *135*
Abell 2199	**63**, 126
Abell 3627	76
Abell 3827	71
Abell 4449	135
Aldebaran	44
Alpha Cru	*111*, 115
Altair	45
AM-4	46
Andromeda Nebula	5, 6, 7, 8, 17, 25, 26, 48, 59, 60, 72, 75, 80, **82**, 106, *120*, 122
Antares	50, *108*
Arktur	80
B 33	*110*, 115
B 72	*111*, 116
B 85	*111*
B 86	*111*, 115, **115**
B 88	*111*
B 89	*111*
B 92	*111*
B 103	*111*
B 133	*111*
B 142	*111*, 116
B 143	*111*, 116
B 168	*111*
B 296	*111*
Barnard's Loop	110
Barnard's Galaxy	61, *121*
Barnard's Star	58
Baxendell's Unphotographable Nebula	35
Bear Paw Galaxy	126, **126**
Beehive Cluster	*104*
Beta And	31, 32, *32*, 62, *120*, 123
Beta CVn	130, *132*
Beta Lyr	53
Big Dipper	44
BL Lac	73, 127, **127**, *128*
Black Eye Galaxy	*121*
Blinking Planetary	*117*, 119
Blue Planetary	*117*
Blue Snowball	38, *118*
Bode's Nebulae	8, *121*, *125*
Brocchi's Cluster	*104*
Bubble Nebula	*111*
Bug Nebula	*117*
Butterfly Cluster	*104*
C 1	34, 103
C 2	34, 117
C 109	34
California Nebula	109, *110*
Canis Major Dwarf	75
Caroline's cluster	101, *104*
Cartwheel	*125*, 126
Castor	9
Cat's Eye Nebula	25, *117*, 119
Centaurus A	67, 125, 126
Centaurus Chain	*126*, 127, *134*, 135
Cetus A	67
Chi Persei	1, 4, 39, **57**, 58, *103*
Christmas Tree Cluster	103
Coalsack	57, *111*, 115
Coathanger	104
Cocoon Nebula	*111*
Collinder 316	101, **102**, *104*

Collinder 399	104	Great Attractor	76
Coma Berenices Cluster	39, 100, 101, 104	Great Rift	57, 115
		HCG 61	**39**, 133
Coma cluster	*121, 135*	Helix Galaxy	*125*
Coma/A1367 supercluster	76, *135*, 136	Helix Nebula	116, *118*
		Hercules cluster	135, *135*
Comet Halley	7	Hercules super-cluster	*135*
Cone Nebula	57, 58, 101, 103, 111, 116		
Copeland Septet	133, *134*	Herschel's Garnet Star	10
Crab Nebula	6, 7, 58, *110*, 113		
Crescent Nebula	*111*	Herschel's Ray	*111*, 114, **114**
Cygnus A	67	Hickson catalogue	133
Delta Cep	69	Hind's Variable Nebula	51, *110*
Deneb	45		
Double Cluster	4, 39, 58, 101, *103*	Hoag's Object	68, **68**, 126, *126*
Double Quasar	128, 129, **129**	Holmberg dwarfs	66
Dreyer, John Louis Emil	100	Homunculus Nebula	*111*, 116
Dumbbell Nebula	7, 18, 38, 58, *117*	Horsehead Nebula	115
Eagle Nebula	101, *104*	Hubble's Variable Nebula	111
Earth	1, 12, 17, 30, 31, 32, 50, 60, 61, 62, 81	Hubble's Variable Nebula	53
Egg Nebula	54, *118*	Hyades	1, 39, 44, 58, 100, 101, *103*
Emu (dark constellation)	115	Hyadra-Centaurus supercluster	*135*
Enceladus	14		
equatorial mounting	122	Hydra cluster	*135*
		I Zw 1	*128*
Eskimo Nebula	*117*	IC 10	56, *120*, 122
ESO 146-IG5	71	IC 59	*110*, 112, **112**
ESO 29-G21	36	IC 63	*110*, 112, **112**
Eta Car	116	IC 405	*110*
Eta Carinae Nebula	**57**, 58, *111*	IC 434	*110*, 115
		IC 443	*110*
Eta Her	45	IC 972	37, 116, *117*
false comet	102, **102**	IC 1101	71
Flame Nebula	*110*	IC 1276	36, 107, *108*
Flaming Star Nebula	*110*	IC 1296	117, *118*
		IC 1318	*111*
Fornax cluster	*120*, 135, *135*	IC 1454	37, 116, *118*
Fornax System	75, 106, *108*	IC 1613	66
G 1	106, **106**, *108*	IC 1684	*134*
Gamma Cas	*110*, 112, **112**	IC 1685	*134*
Gamma Cygni Nebula	*111*	IC 1688	*134*
		IC 1689	*134*
Gamma Lyr	53	IC 1690	*134*
GCL 45	36	IC 2118	*110*, 115
Gem Cluster	*104*	IC 2157	101, *103*
Ghost of Jupiter	*117*	IC 2233	124, *125*, 126, **126**
GNz-11	72, **73**	IC 2391	5, **5**, 101, *104*

IC 2501	116, *117*, **119**	M 15	46, 105, *109*, 119
IC 2602	*104*	M 16	**57**, 58, 101
IC 3074	65, *65*	M 17	6, 18, 48, **57**, 58, *111*
IC 4625	34	M 18	**57**, 58
IC 4665	*104*	M 19	107, *108*
IC 4677	55, **55**, *117*	M 20	7, 48, 57, 58, *111*
IC 4703	101, *104*	M 21	57, 58, *111*
IC 4715	34	M 22	6, 45, **45**, *108*
IC 4725	100, *104*	M 24	8, 34
IC 4756	*104*	M 25	34, 100
IC 5146	*111*	M 26	**57**, 58
IC 5148	*118*	M 27	7, 18, 38, 53, 55, 116, *117*
IC 5150	*118*	M 29	**57**, 58
Ink Spot	*111*, 115	M 30	18, 86, *109*
Integral Sign Galaxy	124, **124**, *125*	M 31	5, 10, 24, 25, 26, 27, 28, **29**, 35, **35**, 60, 63, 69, 75, 90, 91, 98, 106, *108*, *120*, 122
Intergalactic Wanderer	47, *108*		
IV 1	14	M 32	6, **29**, 75, *120*
Jewel Box	**57**, 58, *104*	M 33	5, 23, **23**, 28, 75, 81, 90, 91, 98, *120*, 122
Jupiter	14, 123	M 35	101, *103*
Kemble, Lucian	104	M 36	58, 101, *103*
Kemble's Cascade	104, 105, *105*	M 37	**57**, 58, 101, *103*
Lagoon Nebula	5, 36, 58, *111*, 116	M 38	**57**, 58, 101
Large Magellanic Cloud (LMC)	5, 100, 109, *110*, *120*	M 39	**57**, 58
		M 40	8
LBN 25	36	M 41	4
Leo cluster	*135*	M 42	5, 24, 25, 48, **57**, 58, *110*
Leo Quartet	133, *134*	M 43	48
Leo Triplet	*132*, 133	M 44	4, 7, **57**, 58
Little Dumbbell	*117*	M 45	7, 36, 47, **57**, 58, 100
Local Arm	58	M 46	101, **102**, *103*, *117*, 119
Local Group	28, 31, 36, 61, 64, 66, 70, 71, 75, 81, 106, *108*, *120*, 121	M 47	8, 101
		M 48	8
Local Supercluster	76, 135, 135	M 50	**57**, 58
M 1	6, 7, **7**, 37, 48, 49, **57**, 58, 94, **94**, *110*, 113, *113*	M 51	7, 22, **22**, 23, 24, 26, **26**, 61, 74, 92, 125, *126*
		M 53	107, **107**, *108*
M 2	109	M 54	106, *108*
M 3	46, *108*	M 55	10, *109*
M 4	47, 106, *108*	M 56	*109*
M 5	*108*	M 57	7, 37, 38, 53, 55, *109*, *117*, **118**
M 6	**57**, 58, 86		
M 7	4, **4**, 39, 86, *104*	M 64	*121*
M 8	5, 36, 48, **57**, 58, *111*	M 65	*132*
M 10	107, *108*	M 66	*132*
M 12	107, *108*	M 67	101
M 13	6, 36, 45, 46, 105, *108*	M 70	7

M 71	48, 107, *109*		80, 81, **82**, 90, 105, 107, 115, 122
M 72	104, *109*		
M 73	8, **8**, 104, *109*	Milky Way objects	57
M 74	61, **61**, 86, *120*		
M 75	106, *109*	Mimas	14
M 76	53, *117*	Mirach	31, 32
M 77	61, 67, 72, *125*	Mirach's Ghost	32, *120*, 123
M 78	49, *110*, 115	Mizar	31
M 79	*108*	Moon	1, 25, 28, 45, 85
M 80	10, 47, *108*	Mrk 205	128, *128*, **128**
M 81	8, 65, 75, 86, *121*, 125, *125*	Mrk 421	*128*
M 82	8, 67, 75, 86, *121*, 125, *125*	Mrk 501	*128*
M 83	6, **6**, 61, 65, *121*	My Cep	10
M 84	*121*, 134, **134**, 135	NGC 1	*131*
M 86	*121*, 134, **134**, 135	NGC 2	*131*
M 87	48, 67, *125*, **134**, 135	NGC 40	33, 34, *117*, 119
M 91	8	NGC 55	32, *120*
M 92	*108*	NGC 100	65, 124, *125*
M 94	23, **23**	NGC 104	105, *108*
M 95	65, *121*	NGC 127	**67**
M 96	*121*	NGC 128	66, 67, **67**
M 97	**front**, 8, 53, *117*, 121	NGC 130	**67**
M 99	*121*	NGC 185	63
M 101	23, **23**, 61, 65, *121*	NGC 188	33, 34, 101, *103*
M 102	8, *121*	NGC 205	8, **29**, 75, *120*
M 103	101	NGC 206	35, **35**, *120*
M 104	65, *121*, 124	NGC 221	*120*
M 105	*121*	NGC 224	*120*
M 106	*121*	NGC 246	*117*
M 108	**front**, 117, *121*, 123	NGC 247	*117*
M 110	75, 101, *120*	NGC 253	11, **11**, 107, *108*, *120*
Magellanic Clouds	4, 16, 46, 47, 65, 66, 75, 80, 109, 122	NGC 281	*110*
		NGC 288	107, *108*
Magellanic Stream	80	NGC 292	*120*
		NGC 315	63
Maia Nebula	34, **34**, 51, 101, *103*, 109	NGC 373	*134*
Markarian Chain	*121*, 134, *134*, **135**	NGC 375	*134*
Mayall II	106, **106**, **108**	NGC 379	*134*
Mel 20	*103*	NGC 380	*134*
Mel 22	*103*	NGC 382	*134*
Mel 25	100, *103*	NGC 383	*134*
Mel 96	*117*, 119	NGC 383 group	*134*, 135
Mel 101	*104*	NGC 384	*134*
Mel 111	39, 100, *104*	NGC 385	*134*
Merope Nebula	34, **34**, 51, 101, *103*, 110, 114	NGC 386	*134*
		NGC 387	*134*
Milky Way	1, **3**, 16, 26, 28, 30, 31, 36, 40, 41, 44, 46, 47, 48, 50, 52, 53, 55, 56, 57, 58, 59, **59**, 60, 71, 72, 75, 76, 77,	NGC 388	*134*
		NGC 404	31, 32, **32**, 62, *120*, 123
		NGC 467	*131*, 133

NGC 470	*131*, 133	NGC 1618	*131*, 133
NGC 474	*131*, 133	NGC 1622	*131*, 133
NGC 494	*134*	NGC 1625	*132*, 133
NGC 495	*134*	NGC 1647	*103*
NGC 496	*134*	NGC 1854	*108*
NGC 498	*134*	NGC 1904	*108*
NGC 501	*134*	NGC 1909	*110*
NGC 503	*134*	NGC 1912	*103*
NGC 504	*134*	NGC 1952	*110*
NGC 507	*134*	NGC 1960	*103*
NGC 507 group	*134*, 135	NGC 1976	*110*
NGC 508	*134*	NGC 1977	*110*
NGC 520	67	NGC 1999	*110*, 115
NGC 581	*103*	NGC 2024	*110*
NGC 598	*120*	NGC 2055	100
NGC 604	*120*	NGC 2068	*110*
NGC 628	*120*	NGC 2070	109, *110*
NGC 650	*117*	NGC 2099	*103*
NGC 651	*117*	NGC 2158	101, *103*
NGC 660	*125*	NGC 2163	52, *110*, 112, **112**
NGC 708	*135*	NGC 2168	*103*
NGC 741	63	NGC 2232	9, **10**, 101, *103*
NGC 750	130, *131*	NGC 2237	*110*
NGC 751	130, *131*	NGC 2242	*117*
NGC 752	*103*	NGC 2244	6, 35
NGC 869	*103*	NGC 2261	53, **53**, *110*
NGC 884	*103*	NGC 2264	14, 101, 103, *111*, 116
NGC 891	48, 56, **61**, 65, *120*, 135	NGC 2287	*103*
NGC 910	*135*	NGC 2323	*103*
NGC 985	68, *125*, 126	NGC 2360	11
NGC 1039	*103*	NGC 2362	101, **101**, *103*
NGC 1049	106, *108*	NGC 2367	43, **43**
NGC 1068	*125*	NGC 2392	*117*, 119
NGC 1275	67, *125*, 135, *135*, 136	NGC 2403	*120*
NGC 1333	*110*	NGC 2404	*120*
NGC 1360	116, **117**	NGC 2419	47, **47**, 106, *108*
NGC 1365	*120*	NGC 2422	*103*
NGC 1399	*135*	NGC 2437	*103*
NGC 1432	34, 51, 101, 109	NGC 2438	101, **102**, *103*, *117*, 119
NGC 1435	34, 51, 101, *110*, 114	NGC 2537	*125*, 126, **126**
NGC 1454	37	NGC 2548	*103*
NGC 1499	109, *110*	NGC 2573	*120*, 122
NGC 1502	104, 105, **105**	NGC 2623	125, 126
NGC 1514	18, 19, **19**, *117*, 119	NGC 2632	*104*
NGC 1535	*117*	NGC 2682	*104*
NGC 1554	51	NGC 2685	*125*
NGC 1555	51, **51**, *110*	NGC 2689	24
NGC 1600	63	NGC 2736	*111*, 113, 114, *114*

NGC 2818	*117*, 119	NGC 4206	*132*, 133
NGC 2818A	119	NGC 4216	*132*, 133
NGC 2843	14, **14**	NGC 4222	*132*, 133
NGC 2903	*120*	NGC 4242	65
NGC 2905	*120*	NGC 4254	*121*
NGC 3031	*121*	NGC 4258	*121*
NGC 3034	*125*	NGC 4291	128, *128*
NGC 3073	**129**	NGC 4298	130, *132*
NGC 3079	129, **129**	NGC 4302	130, *132*
NGC 3115	63, *121*	NGC 4319	128, *128*
NGC 3172	*121*, 122	NGC 4361	53, **54**, *117*
NGC 3185	*134*	NGC 4365	63
NGC 3187	*134*	NGC 4371	66
NGC 3190	*134*	NGC 4374	*121*
NGC 3193	*134*	NGC 4406	*121*
NGC 3195	34	NGC 4435	*134*, **134**
NGC 3242	*117*	NGC 4438	*134*, **134**
NGC 3293	*104*	NGC 4449	32, 65, 66, **66**, 125, *125*
NGC 3309	*135*	NGC 4458	*134*, 134
NGC 3311	*135*	NGC 4461	*134*, **134**
NGC 3314	74, **74**, *121*	NGC 4473	*134*, **134**
NGC 3351	*121*	NGC 4477	*134*, **134**
NGC 3368	*121*	NGC 4479	*134*, **134**
NGC 3372	*111*	NGC 4485	130, *132*
NGC 3379	*121*	NGC 4486	*125*
NGC 3384	*121*	NGC 4490	130, *132*
NGC 3389	*121*	NGC 4565	61, 65, *121*
NGC 3556	*121*	NGC 4567	130, **131**, *132*
NGC 3585	62, **63**	NGC 4568	130, **131**, *132*
NGC 3587	*117*	NGC 4594	*121*
NGC 3628	*132*	NGC 4622	*134*
NGC 3745	*134*	NGC 4622B	*134*
NGC 3746	*134*	NGC 4627	*132*, 133, **133**
NGC 3748	*134*	NGC 4631	*132*, 133, **133**
NGC 3750	*134*	NGC 4650	*134*
NGC 3751	*134*	NGC 4650A	*126*, 127, *134*
NGC 3753	*134*	NGC 4656	*132*, 133, **133**
NGC 3754	*134*	NGC 4657	*132*, 133, **133**
NGC 3842	*135*	NGC 4676	130
NGC 3918	*117*, 119, **119**	NGC 4676A	*132*
NGC 4038	**81**, 130, *132*	NGC 4676B	*132*
NGC 4039	**81**, 130, *132*	NGC 4715	*134*
NGC 4111	65	NGC 4755	101, *104*
NGC 4151	72	NGC 4826	*121*
NGC 4169	**39**, *134*	NGC 4861	68, *126*
NGC 4173	**39**, *134*	NGC 4874	*135*, 136
NGC 4174	**39**, *134*	NGC 4879	*135*
NGC 4175	**39**, *134*	NGC 4889	*121*, 136

NGC 5023	*126*	NGC 6720	*117*
NGC 5024	*108*	NGC 6723	*111*
NGC 5053	107, **107**, *108*	NGC 6726	*111*, 114, **114**, 115
NGC 5128	63, 67, 125, *126*	NGC 6727	*111*, 114, **114**, 115
NGC 5139	*108*	NGC 6729	*111*, 111, 114, **114**, 115
NGC 5194	*126*	NGC 6745	126, *126*
NGC 5195	74, 125, *126*	NGC 6779	*109*
NGC 5236	*121*	NGC 6809	*109*
NGC 5272	*108*	NGC 6822	61, *121*
NGC 5457	*121*	NGC 6826	*117*, 119
NGC 5466	47, **48**, 106, *108*	NGC 6838	*109*
NGC 5529	65	NGC 6853	*117*
NGC 5544	130, *132*	NGC 6864	*109*
NGC 5545	130, *132*	NGC 6888	*111*
NGC 5866	*121*	NGC 6946	121, 122, **122**
NGC 5904	*108*	NGC 6960	99, **99**
NGC 5907	61, 65, *121*, 123, 123, 124	NGC 6981	*109*
NGC 6027	*134*	NGC 6992	18, **18**, *111*
NGC 6042	*135*	NGC 6995	*111*
NGC 6093	*108*	NGC 7000	57, 58, 109, *111*
NGC 6121	*108*	NGC 7006	47, 106, *109*
NGC 6166	**63**, 64, 126, *126*	NGC 7008	*117*
NGC 6205	*108*	NGC 7009	10, 14, **48**, *118*, 119
NGC 6207	*108*	NGC 7026	116, *118*
NGC 6210	*117*, 119	NGC 7027	*118*
NGC 6218	*108*	NGC 7078	*109*
NGC 6231	101, **102**, *104*	NGC 7088	35, **35**
NGC 6240	125, *126*	NGC 7089	*109*
NGC 6254	*108*	NGC 7092	*104*
NGC 6273	*108*	NGC 7099	*109*
NGC 6302	*117*	NGC 7252	126, *126*
NGC 6341	*108*	NGC 7293	116, *118*
NGC 6405	*104*	NGC 7317	*134*
NGC 6475	*104*	NGC 7318A	*134*
NGC 6514	*111*	NGC 7318B	*134*
NGC 6520	*115*	NGC 7319	*134*
NGC 6523	*111*	NGC 7320	133, *134*
NGC 6543	80, **55**, *117*, 119	NGC 7320C	133, *134*
NGC 6543 (spectrum)	**25**	NGC 7331	*121*, 133
		NGC 7332	66, 130, **130**, *132*
NGC 6544	107, *108*	NGC 7335	130, **130**
NGC 6611	*104*	NGC 7339	**130**, *132*
NGC 6618	*111*	NGC 7479	96, **96**, *121*
NGC 6656	*108*	NGC 7635	*111*
NGC 6702	123	NGC 7640	65
NGC 6705	*104*	NGC 7654	*104*
NGC 6715	*108*	NGC 7662	38, *118*
NGC 6717	36, 37, **37**, 107, *109*	NGC 7743	65

NGC 7789	101, *104*	Ruprecht 106	46
NGC 7793	65	S 106	111
North America Nebula	58, 109, *111*	S5 0014+81	127, 128
		Sagittarius Arm	**57**, 58
Ny Eri	*131*, 133	Sagittarius Dwarf	75, 80, *108*
Ny1 Sgr	107, *108*	Saturn	**7**, 17, 45, 123
Ny2 Sgr	37, **37**, 107, 108	Saturn Nebula	10, *118*
Oberon	12	Sculptor Galaxy	*120*
OCL 421	36	Sculptor System	64, **64**, 75
Omega Centauri	6, 46, 105, *108*	Scutum-Centaurus Arm	58
Omega Nebula	6, 18, 58, *111*		
Omikron Velorum Cluster	**5**, 39, *104*	Segue 2	71, 75
		Seyfert Sextet	133, 134
Orion Arm	**57**, 58	Sgr A	60
Orion Nebula	frontispiece, 5, 7, 10, 17, 18, 22, 23, 25, 26, 48, 49, 51, 58, *110*	Sgr A*	60
		Sh2-25	36
		Sh2-106	52, **52**, *111*
Outer Arm	58	Shakh 1	75, **75**
Owl Nebula	8, 58, *117*	Siamese Twins	130, **131**, *132*
Palomar 7	36, 107, *108*	Silver Dollar	*120*
Palomar 9	36, 37, **37**, 107, *109*	Sirius	4, *103*
Palomar 12	46	Small Magellanic Cloud (SMC)	5, 36, 100, 105, *120*
Pease 1	*109*, 119		
Pencil Nebula	*111*, 113, 114, **114**	Snake Nebula	*111*, 116
Perseus A	67, *136*	SNR 184.6-05.8	37
Perseus Arm	**57**, 58	Sombrero Galaxy	*121*, 124
Perseus Cluster	101, *103*, 135	Southern Cross	45
Perseus cluster (galaxies)	*125*	Southern Pleiades	*104*
		Spindle Galaxy	63, *121*
Perseus-Pisces supercluster	*135*	Stephan's Quintet	*121*, 133, 134
		Struve's Lost Nebula	51
PGC 28990	**129**		
Pillars of Creation	101	Summer Triangle	45
Pinwheel Galaxy	*121*	T Tau	44, 51, **51**, *110*
PK 63-13.1	37	Tarantula Nebula	109, *110*
Pleiades	1, 7, 18, 34, **34**, 39, 51, 58, 90, 100, 101, *103*, *110*, 114	Tau CMa	101, **101**, *103*
		The Antennae	**81**, 130, *132*
PN G063.1+13.9	38	The Box	**39**, 133, *134*
Polaris	9	The Mice	130, *132*
Polarissima Australis	*120*, 122	Theta Ori	10, *110*
		Titania	12
Polarissima Borealis	*121*, 122	Trapezium	10, *110*
		Triangulum Nebula	5, *120*
Praesepe	1, 4, 7, 39, 58, *104*		
Ptolemy's Cluster	**4**, 86, *104*	Trifid Nebula	7, 58, *111*, 116
Q 0957+561	*128*, 129, **129**	Trümpler 24	101, **102**, *104*
R CrA	114, **114**, 115	UGC 2885	123, **123**
R Mon	53, **53**	UGC 3697	124, **124**
Ring Nebula	7, 38, 53, 58, *117*, 118, **118**	UGC 3714	**124**
Rosette Nebula	**57**, 58, *110*		

UGC 5938	68, **68**
UGC 5942	68, **68**
UGC 12591	64
Uranus	9, 11, 12, 14, 53, *117*, 119
Ursa Major moving group	44
Vega	26, 45
Veil Nebula	18, **57**, 58, *111*, 113
Vela supernova remnant	113, 114
Venus	32
Virgo A	67
Virgo Cluster	8, 28, 65, 70, 76, *121*, *125*, 125, 130, *132*, *134*, *135*, 135
W Com	127, *128*
Whale Galaxy	*132*
Whirlpool Nebula	7, 22, **22**, 23, 26, *126*
Wild Duck Cluster	101, *104*
Witch Head Nebula	*110*, 115
Zeta Her	45
Zeta Tau	7
Zeta1 Sco	101, 102, **102**
Zeta2 Sco	101, 102, **102**

Sources of Figures

l = left figure, *r* = light figure, *up* = upper figure, *lo* = lower figure

Digital Sky Survey: 8, 10, 19 *r*, 35 *lo*, 37 *l*, 43, 63 *up*, 64, 101, 102 *lo*, 105, 119 *up*, 119 *lo*, 124

Hubble Space Telescope: 68 *up*, 73, 74, 80, 106, 127 *lo*, 129 *inset*

NASA: 59, 82

Sloan Digital Sky Survey: front cover, 14, 28, 32, 39, 47, 48 *l*, 48 *r*, 65, 67, 68 *lo*, 75, 107, 123 *up*, 129

Stefan Binnewies, Josef Pöpsel (Capella-Observatory): 4, 5, 6, 7, 29, 37 *r*, 45, 51, 52, 53, 54, 55, 61 *l*, 61 *r*, 63 *lo*, 66, 81, 99, 112 *up*, 113, 114 *up*, 114 *lo*, 115, 118, 122, 123 *lo*, 131, 133, 134, 136, back cover

Virgo Consortium: 78

Wolfgang Steinicke: frontispiece, 3, 9, 11, 12, 15, 16, 18, 19 *l*, 20, 21, 22, 23, 25, 26, 27, 34, 35 *up*, 42, 56, 57, 62, 66 *inset*, 83, 87, 89, 94, 96, 102 *up*, 112 *lo*, 126, 127 *up*, 128, 130, appendix

Dr Wolfgang Steinicke studied physics and mathematics in Germany. He specialised in General Relativity and Quantum Field Theory. Already in his youth, he observed the sky with telescopes. Later his interest focused on Dreyer's *New General Catalogue*, which essentially rests upon observations by William and John Herschel. The research on non-stellar objects, their data and historical sources led to comprehensive catalogues, including a revision of the NGC and its supplements. In 2008, he received a PhD at Hamburg University with a thesis on nineteenth century deep-sky observations, published 2010 by Cambridge University Press as *Observing and Cataloguing Nebulae and Star Clusters: From Herschel to Dreyer's New General Catalogue.*

Steinicke is a Fellow of the *Royal Astronomical Society*, director of the History of Astronomy section of the German *Vereinigung der Sternfreunde*, committee member of the British *Webb Deep Sky Society* and works for international associations. He frequently organizes astronomy meetings and gives talks or courses all over the world. Steinicke is the author of nine books (German and English) and has published more than 300 scientific articles. Currently he writes a comprehensive book about William Herschel and his observations.

www.klima-luft.de/steinicke
steinicke@klima-luft.de

www.ingramcontent.com/pod-product-compliance
Lightning Source LLC
Chambersburg PA
CBHW050100230526
45470CB00004B/1617